FOLK ILLUSIONS

FOLK ILLUSIONS

Children, Folklore, and
Sciences of Perception

K. Brandon Barker and Claiborne Rice

INDIANA UNIVERSITY PRESS

This book is a publication of

Indiana University Press
Office of Scholarly Publishing
Herman B Wells Library 350
1320 East 10th Street
Bloomington, Indiana 47405 USA

iupress.indiana.edu

Manufactured in the United States of America

Cataloging information is available from the Library of Congress.

ISBN 978-0-253-04108-1 (hardback)
ISBN 978-0-253-04109-8 (paperback)
ISBN 978-0-253-04110-4 (ebook)

1 2 3 4 5   24 23 22 21 20 19

To our families

CONTENTS

PREFACE

Zane's Illusion

G O LIKE THIS. Hold your right hand out in front of you so that your arm extends perpendicularly from the center of your chest. Your right palm should be facing up so that you can see the underside of your right forearm. Now, make a fist with your right hand. The next step is simple. Pull your arm up in front of your face so that the distal portion of your forearm touches your nose and the knuckles of your fist point toward the sky. From this position (you can do this while you read), point the index finger on your left hand in the standard *pointing position*. Now, hold that pointed finger so that it is parallel to the ground and pointing to your right. Here comes the fun part. While staring straight ahead, slowly pass your pointed finger in front of your right forearm that is in front of your face.

Zane's mother, Rose, first relayed Zane's narrative description of the trick's origins in the spring of 2017. That semester, Rose was enrolled in Brandon Barker's Children's Folklore course at Indiana University. She developed a habit of showing Zane, her nine-year-old son, some of the games, rhymes, and other activities that she learned about in class. Then one evening over dinner, Zane showed his mother the shrinking/disappearing-finger illusion we described above. *What an excellent trick!* Correctly, Rose recognized that we would be interested to learn about the trick, so she asked Zane where he had learned it. His story, Rose told us, went something like this: Zane—while sitting in school one day—was bored and discouraged. Being "strapped" to a school desk can be torturous for children who value playful, fully embodied experience over pencils and paper. In this state of mind, Zane sort of "found himself" in the position that facilitates Zane's illusion as he, in frustration, rested his forehead in his palm. From there, Zane told his mother, he simply fiddled with his fingers until he "discovered" the visual illusion.

It is an appealing tale—both a testament to a bored nine-year-old's ingenuity and a reminder that children *do not* need extravagant toys or electronic charms in order to entertain themselves. But as we said, the story

Fig. o.1. Zane performs Zane's illusion at his kitchen table.

is more complicated than just this. In a fortuitous schedule overlap, Zane had a free day to visit our Children's Folklore course the following Monday because his school was closed for the observance of Presidents' Day. On that Monday, we were discussing Yo-Mama joke cycles and toast-like battling traditions among children and teens. Much of youthful folklore, and thus much of a course on children's folklore, pushes the boundaries of acceptability. Children possess a keen knack for walking all over taboos. As a true consultant, Zane was prompted to share with us a Yo-Mama joke. After a sly, careful glance at his mother, he delighted the room: "Yo mama is so ugly, she tried to enter an Ugly Contest, and the judges said, 'Sorry—no professionals!'" Rose's fellow students burst into cacophonous laughter. We had all just broken the rules by allowing her nine-year-old to perform for us what a child is "supposed" to perform only for his peers. It was a great pleasure and, clearly, Zane was a star.

Feeding off of this success, Brandon explained to Zane that the class was also very happy to have learned about Zane's illusion and that they had all been impressed with the trick. Sensing the moment was right, Rose then asked Zane to re-create his tale of discovering Zane's illusion:

"Mom, I don't know what you're talking about." Zane blinked up at his mother.

"Remember, at the dinner table, you told me about how you were frustrated at school, and how . . ." Tilting her head, Rose tried to jog Zane's memory.

Blink . . . Blink . . . "I never said that, Mom. I don't know what you're talking about."

Rose, surprised and a little embarrassed, laughed at Zane's apparent amnesia. As the conversation moved on, the rest of us chalked Zane's memory lapse up to the precarious mind of children. Brandon consoled her: "They'll do that to you."

It turns out, Zane's illusion is only one of the many forms of children's play that distort the boundaries between illusion and reality. In the pages and chapters that follow, we will outline our understanding of these kinds of play by examining many such forms, and we will argue that these forms make up a discrete genre of folklore. In chapters 1 and 2, we outline the central terms of our work and the pervasive components of these forms. Chapters 3 and 4 consider the active processes of transmission and embodied perception as mutually constitutive. Chapter 5 presents a case study of children's play with weight illusions, and chapter 6 problematizes the boundaries of our genre in the context of a well-known form of children's supernatural folklore: mirror summonings. Our concluding chapter outlines our notion of body acquisition, after which we provide an appendix of all the illusions that children, youths, and young adults have taught us. (Zane's illusion is B14.)

The appendix—though it is situated at the end of the book—is a good place to start. More than once, we will recommend taking a look at it before completing the chapters that precede it. And we do so now if for no other reason than to remind you that you too once performed Zane's illusion–like activities. If you are thinking that maybe you have never performed a trick like Zane's illusion, let us start by cautioning you: What is true of perception is true for memory. *Assumptions carry great risks.*

ACKNOWLEDGMENTS

WE ARE INDEBTED TO MANY PEOPLE FOR VARIOUS kinds of contributions to this book. First and foremost, we are thankful to the children and youths who have taught us how to play with illusions. We thank the teachers and administrators at St. Cecilia Middle School in Broussard, Louisiana, and at Parents Day Out Preschool in Bloomington, Indiana, for allowing us to break up the routines of their days. Similarly, we thank the Evangeline Area Council, Boy Scouts of America, for their willingness to speak about the illusions their young Scouts perform. We thank Cheryl Hesse for hosting our very first opportunity to observe kids performing folk illusions at her beautiful and hospitable home in Lafayette, Louisiana. We are grateful to several youthful star performers: Aoife, Lucas, Monica, Sofia, Jacob, Calvin, Finley, Zane, Addy, and Lilly. And to the students at Indiana University and the University of Louisiana who have helped us fill out our catalog, we say thank you.

We have benefited from the rich intellectual support of the Department of English at the University of Louisiana and the Department of Folklore and Ethnomusicology at Indiana University. In Louisiana, we thank, especially, our friends and colleagues John Laudun, Barry Ancelet, Marcia Gaudet, Shelly Ingram, Skip Fox, Jonathan Goodwin, Mark Honegger, Wilbur Bennett, Yung-Hsing Wu, Jerry McGuire, Marthe Reed, Elizabeth Bobo, Carmen Comeaux, Keith Dorwick, and many others whose feedback and reinforcement have been invaluable. In Indiana, we developed enriching relationships with Diane Goldstein, Pravina Shukla, Gregory Schrempp, John McDowell, Jason Jackson, Ray Cashman, David McDonald, Tim Lloyd, Sue Thuoy, Fernando Orejuela, Daniel Reed, Alan Burdette, Ruth Stone, Rebecca Dirksen, and Alisha Jones. Talking for hours and hours with our friend Daniel Povinelli has sharpened both our thinking and our work with folk illusions. The same is true for our friend Henry Glassie, who, over many cups of coffee, has guided and encouraged this project, and we thank him for reading an early draft.

Outside of Louisiana and Indiana, we received tremendous aid from researchers in folklore and various other fields as we tried to avoid oversimplifying the important contributions that other disciplines have made to the study of folk illusions. We thank, especially, Elizabeth Tucker, Katharine Young, Susana Martinez-Conde, Stephen Macknik, Jack De Havas, Giovanni Caputo, Mike Kalish, Fabrizio Benedetti, Elliot Oring, and Nina Fales.

A succession of editors have urged us forward responsibly while helping our work to appear in its best form. We thank Jim Leary, Tom DuBois, Michael Dylan Foster, and Jon Sutton. At Indiana University Press, our project has grown and improved with the help of Gary Dunham, Janice E. Frisch, and Kate Schramm.

It is impossible to thank all of the many people who agreed to talk with us about their own experiences of folk illusions, either briefly or at length. We have had extended discussions over many years with several of our friends and students that have yielded not only examples of specific folk illusions, but also insights into the genre as a whole. This group includes Tim Henson, Nik Norrod, Samuel Martin, Micah Loewer, Mike Quinn, Tina Mitchell, Amanda LaRoche, Prasanna (Monty) Kawatkar, Wayne Arnold, Emma Tomingas, Matt Sackmann, Erin Holden, Angela Granese, Corey Green, Charmika Stewart, Elizabeth Underwood, Brendon Vayo, and Bryan Hinojosa.

Finally, we thank our families because of all the people with whom we have exchanged discussion, remembrances, and demonstrations, family members have spent the most time as, variously, informants, observers, interlocutors, test subjects, and philosophical debaters. Brandon's wife, Ellen, is an educator of children, which means that she is a true expert of the child's ethnic point of view. Brandon and Ellen's little Zoa Kyoko is still willing to play Peekaboo from time to time, and their second daughter, Rosaline Fern, will soon join in the fun. Ken, Cindy, Shawn, Shaylyn, and Kelsie have been a part of it all along. Clai's sons Sam Whitt and Will Rice were teenagers at the inception of this study, so they provided many new forms and also were willing to try out illusions that we discovered elsewhere. Clai's wife, Lydia, has a great memory for her childhood, and as a scholar of American literature was a fountain of literary and critical references for both specific activities and for many of the themes and ideas throughout the book. She also read and reread many chapter drafts. We are grateful for all of their patience as this book wended its way toward completion.

ACCESSING AUDIOVISUAL MATERIALS

AUDIOVISUAL MATERIALS ARE AVAILABLE FOR THIS VOLUME AND can be viewed online via the Indiana University Media Collections Online at https://purl.dlib.indiana.edu/iudl/media/s55m312x7z. Information and links for each individual entry follow.

Video 1. Zoa Plays Peekaboo. https://purl.dlib.indiana.edu/iudl/media /z70890t32b.

Video 2. A2, Where Am I Touching You? https://purl.dlib.indiana.edu/iudl /media/890r76g668.

Video 3. A3, The Chills. https://purl.dlib.indiana.edu/iudl /media/494v53m07m.

Video 4. A5, Dead Man's Hand. https://purl.dlib.indiana.edu/iudl/media /h24673271f.

Video 5. A6, Touching Invisible Glass. https://purl.dlib.indiana.edu/iudl /media/197x61fp47.

Video 6. A7, Furry Air. https://purl.dlib.indiana.edu/iudl/media/d56z904326.

Video 7. A8, Electricity at the Fingertips. https://purl.dlib.indiana.edu/iudl /media/8515645w9f.

Video 8. A11, String Pull. https://purl.dlib.indiana.edu/iudl/media/188158cj1f.

Video 9. B13, Floating Finger. https://purl.dlib.indiana.edu/iudl /media/277386198h.

Video 10. B14, Zane's Illusion. https://purl.dlib.indiana.edu/iudl/media /b78t34tp83.

Video 11. C1, Buzzing Bee. https://purl.dlib.indiana.edu/iudl/media /b29b19gb5h.

Video 12. C2, The Church Bell. https://purl.dlib.indiana.edu/iudl /media/910j82mc59.

Video 13. D1, Winding-Cranking Fingers. https://purl.dlib.indiana.edu/iudl /media/b29b19gb46.

Video 14. D2, Magnetic Rocks. https://purl.dlib.indiana.edu/iudl/media /g54x51jr18.

Video 15. E3, I Can't Move My Finger. https://purl.dlib.indiana.edu/iudl /media/q77f75sk2t.

Video 16. E4, Twisted Hands. https://purl.dlib.indiana.edu/iudl /media/603q87d079.

Video 17. G1, Falling through the Floor. https://purl.dlib.indiana.edu/iudl /media/207t64hq95.

FOLK ILLUSIONS

1

EVERYONE KNOWS THAT SEEING IS
(*NOT ALWAYS*) BELIEVING

L IKE MOST SCHOOLS, ST. CECILIA CATHOLIC MIDDLE SCHOOL in Broussard, Louisiana, educates its youth according to a system of time slots. Students gather in assigned classes for daily lessons: English and language in the morning. Then history, then religion. Geography and science in the afternoon. Teachers present regimented educational programs around necessary breaks for eating, for Mass, and for some free play. The playgrounds offer release for nine-, ten-, eleven-, and twelve-year-old girls and boys who greatly desire time and space to live in the moment, free from the weight of learning objectives. But then again, those of us who remember middle school well know that play is not confined to the playground. Who doesn't recall silly, handwritten notes passed around the classroom? Or the reshaping of homework assignments into aeronautical paper masterpieces? Or weirdly contorted faces of mockery aimed at the teacher's back? Play finds its way into most aspects of youthful school days.

Less obvious to the energetic children and to their adult authorities is the inverse truth that education is not confined to books and lesson plans. A good deal of what youths come to know about the world, about each other, and about themselves is learned during the kinds of social interactions that we adults usually call play. We observed an excellent example of this fact at St. Cecilia in the spring of 2011, just before Ms. Hesse's eighth-grade science class. The bell had just rung, and students were settling in for their daily lesson on the Milky Way. Plastic models of Earth, Venus, Mars, Jupiter, and the other planets hung from the ceiling. Since Ms. Hesse's room was set up for scientific experimentation, the room held no school desks. Students instead sat facing each other across rectangular tables. One boy, who was sitting in front of his friend, held his pencil between his thumb and his index finger. He held it at eye level, about a foot from his face, and began to wiggle the pencil up and down, over and over. The boy was performing the Rubber Pencil illusion. It's possible that several weeks earlier, his classmate sitting just in front of him had shown him how to hold his pencil this way. Maybe his classmate had also shown him how to wiggle his pencil so that it appeared to bend as if made of rubber. Regardless of how the boys learned the trick, neither looked away from this rigid pencil magically and inexplicably bending up and down right in front of their eyes. The boys only stopped staring at the illusion when Ms. Hesse instructed the students to get out their textbooks and to turn to the chapter on our solar system.

From a more scientific point of view, this is what those boys "demonstrated" as they performed Rubber Pencil that day: *When humans observe a rigid rod undergoing simultaneous translational and rotational motion in an arc of approximately 110° and at a frequency of approximately 2.5–3 cycles-per-second, humans perceive that rod to be elastic as if made of rubber.* The first experimental study of Rubber Pencil was not published until 1983, by James Pomerantz. Pomerantz, who clearly understood the folkloric nature of the activity, begins his article with a reference to the tradition's popularity: "The 'rubber pencil illusion' . . . is a striking visual phenomenon that has gone unnoted in the scientific literature on human visual perception, despite its familiarity to many laypeople" (1893, 365).[1] In Pomerantz's analysis, the Rubber Pencil illusion occurs due to the nature of the afterimages that appear during a human's visual perception of moving objects. The densest afterimages occur in the middle area of the pencil, and the least dense afterimages occur at the pencil's ends. These varying densities produce a unified perception of bending.

Pomerantz does not tell us about how he first came to discover the Rubber Pencil illusion, so we are left to guess whether or not some childhood friend demonstrated the activity to a younger version of the scientist. He does add, though, that stage magicians, who have long performed the trick, *do not* seem to understand the "scientific" principles behind the illusion. The scientist quotes George Gilbert and Wendy Rydell—authors of *Great Tricks of the Master Magicians* (1976)—who describe the Rubber Pencil illusion in this fashion: "There's no trick involved here; it just works that way" (131).

Like stage magicians, Ms. Hesse's students could not supply a mathematical understanding of translational and rotational arc degrees or cycles per second attendant to a successful performance of Rubber Pencil, but Ms. Hesse's students' successful performances do raise the following question:

Fig. 1.1. Five-year-old Calvin attempts to perform Rubber Pencil.

What *does* the child performing Rubber Pencil know about the involved illusory perceptions? We have to admit that Ms. Hesse's student successfully performed the required "translational" and "rotational" movements at proper intervals. We have to admit the boy successfully turned his rigid No. 2 pencil into an illusory, rubbery substance. His actions and his engaged classmate's wonderment serve as proof.

Scientific experimentation and mathematical description aside, Ms. Hesse's childish philosophers examined eternal questions about reality as they performed the Rubber Pencil illusion. Experimental evidence suggests that infants as young as one or two months can recognize the difference between rigid and malleable objects, and there's no doubt that Ms. Hesse's eighth graders possessed a well-developed understanding of rigidity.[2] Most children who perform Rubber Pencil probably understand a good deal about the reality of the situation (i.e., the fact that the pencil is not turning from wood to rubber to wood). In 2014, we played Rubber Pencil with a group of seven-year-olds at a summertime day care in Bloomington, Indiana. As the children performed the trick, one seven-year-old yelled out, "It's an optical illusion!" More than questioning the reality of the pencil, the Rubber Pencil illusion prompts introspective questions about the nature of the self.

Seeing—in this case—is not believing; visual perception—like rubber—is malleable. The Rubber Pencil illusion is a lesson as much as it is play.[3]

As a cultural tradition passed along via word of mouth from child to child for generation after generation, the Rubber Pencil illusion is also folklore. The illusion saturates the communal understanding of visual perception for any group in which it is performed. When a child shows another child Rubber Pencil, the children do more than recognize an illusory tendency in their embodied perception: They recognize that the illusions can be produced for each other. They recognize that the experience can be intersubjective, that sharing the illusion can be fun. When the illusion is shown to someone new, the children play with the fact of knowing something about the new performer's bodily perceptions that were up until that point completely unknown to that individual. The outsider to the trick turns insider before the look of astonishment disappears from her face. *Spreading the word*, knowledge begets play begets knowledge begets play.

Folk Illusions

The goal of this book is to identify and examine the shared aspects of embodied perception that children, adolescents, and occasionally adults communicate as they play with perception's illusory qualities. It turns out that most people with whom we have worked know and have performed several activities like Rubber Pencil. Taken together, this group of activities constitutes a genre of folklore we have named *folk illusions*. We define folk illusions as traditionalized verbal and/or kinesthetic actions performed in order to effect an intended perceptual illusion for one or more participants.

Folk illusions present an excellent opportunity for the situated, socially contextualized study of perception during the marked years of social maturation and physical development, and while our methods are largely folkloristic, our search for the ways that culture affects the human body and embodied experience remains inherently interdisciplinary. A full understanding of folk illusions requires two generally separate kinds of data sets: folkloristic methods for gathering cultural traditions in the field *and* experimental methods for identifying and highlighting particular embodied processes (or systems of processes) attendant to any given intended illusion. Our work is folkloristic, but we could not have come to our understanding of folk illusions without taking seriously the controlled data sets that follow more traditional scientific experimentation alongside the data sets that can only be identified out in the real world—outside the confines of the lab.

We have spent eight years employing folkloristic techniques for observing performances and gathering remembrances of types of play that create a perceptual illusion for one or more participants. We have worked with college students, middle schools, Boy Scout troops, and preschools. We have organized fieldwork observations with groups as small as four children, and we have surveyed college courses with as many as three hundred late-teenage, early-adult students. We have found that many people can provide an example of a form of play that creates a perceptual illusion. Most of our work has taken place in two geographical areas of the United States: the Acadiana region of Louisiana and the hills of southern Indiana. Because we often work with university students, we have been able to gather some remembrances of folk illusions from twenty-one states and from nine countries on four continents. One ancillary goal of this book is to promote more fieldwork of folk illusions by scholars working in other languages and cultures.

Why Folklore? Why Illusions?

In folklore and illusions, our work brings together two long-standing subjects of philosophical, scholarly interest. Professional folklorists cite the birth of their discipline in the works of nineteenth-century German philosophers and cultural linguists like Johann Gottfried Herder (1744–1803) and the well-known brothers Jacob (1785–1863) and Wilhelm Grimm (1786–1859), and, of course, the systematic analysis of illusory perception permeates several philosophical traditions from antiquity to the present. At this point, however, we are not aware of any other folkloristic research program focusing on perceptual illusions, and while it is true that psychologists, cognitive scientists, anthropologists, and others have worked at the intersection of perceptual illusions and culture—especially on the problems of cultural influence—we know of no works in those disciplines that focus on perceptual illusions in the context of situated cultural traditions, in the context of folklore.

Along with the older subjects of folklore and illusions, our project also grows out of a relatively recent surge of scholarly attention to questions of embodiment. Indeed, much discussion about what it means to be human has turned toward discussions of human embodiment. In general, the term *embodiment* signifies the physical manifestation of an entity or an essence in a discrete material form, and in the context of humanness, embodiment can loosely be understood as the manifestation of everything that it is to be a human within the boundaries of the material makeup and (inter)actions of that human body.[4]

At this point, the diversity of disciplines dealing with questions of embodiment has given rise to an equally diverse and impressively large literature on the subject. In recent decades, philosophers, cognitive scientists, and neuroscientists have attempted to isolate myriad mental processes of brain and body (Lakoff and Johnson 1999; Gibbs 2005). Linguists—who have long studied the physiological processes of phonetics and audition—have set out to ground pragmatics, semantics, and even grammar in the body and the body's interaction with the physical world (Goodwin 2000; Johnson [1987] 1990; Bergen and Chang 2005). Using theories of embodiment as a governing paradigm for theoretical analysis, anthropologists have started dissecting cross-culturally realized aspects of human life, such as religion, medicine, and food (Csordas 1990, 1994a, 1994b; Mol 2008; Sutton 2001). Clinicians are reconsidering psychological pathologies like anorexia, phantom limbs, and apraxia as psychophysical symptoms of mismatched body maps and body schemas (Ramachandran and Blakeslee 1999; Maclachlan 2004). Folklorists have coined the term *bodylore* in reference to the study of the ways that cultural traditions act upon, alongside, and within the body (Young 1993; Sklar 1994, 2005).

One theme that runs consistently through cross-disciplinary studies of embodiment is the resounding call for a much-needed correction of older dualist, mind-body paradigms associated with philosophers like Plato and Descartes. Drawing especially from the philosophy of phenomenologist Maurice Merleau-Ponty, many of these studies argue instead for viewing the body as inexorably bound to the most abstract and ethereal aspects of the mind. Psychologist Raymond Gibbs describes embodiment's place in contemporary sciences of mind as a necessary, ultimate ground: "People's subjective, felt experiences of their bodies in action provide part of the fundamental grounding for language and thought. Cognition is what occurs when the body engages the physical, cultural world and must be studied in terms of the dynamical interactions between people and the environment. Human language and thought emerge from recurring patterns of embodied activity that constrain ongoing intelligent behavior. We must not assume cognition to be purely internal, symbolic, computational, and disembodied, but seek out the gross and detailed ways that language and thought are inextricably shaped by embodied action" (2005, 9). This is the first, and most important, way that theories of embodiment beneficially coincide with contemporary folkloristics' methodological and theoretical leanings, for the best twentieth-century performance theories in folkloristics and other social sciences moved away from Cartesian dualism as much as they moved away from nineteenth-century textualism. Beginning with the premise that human minds are embodied, we deny—once and for all—abstractions that do not consider the subjective (and intersubjective), phenomenological experience of the folk during a performance. We deny the notion that folklore exists extracorporeally, like so many diffused Grimms' tales hopping along from place to place on legs of their own. We ground performance in its attendant corporeal processes.

As a genre, folk illusions compel us to make central the phenomenological character of traditional performances that induce illusory percepts. Our work catalogs a number of folkloric traditions that have not been previously cataloged. For folklorists, to whom cultural diversity is certainly just as important as biological universals, the expansion of our catalogs remains an imperative descriptive task in and of itself. As of the time of this writing, we have identified seventy discrete folk illusions and another thirty or so variants. You can find our catalog of these folk illusions in the appendix at the back of this book. (It may be helpful to examine—even briefly—our catalog as a way of acquainting yourself with the genre.) Let us add, though, that we are still on the hunt for new folk illusions performed inside and outside of the folk groups we have worked with so far. Our catalog continues to grow.

Our theoretical concerns move past empirical addition to the catalogs of bodylore and folkloristics because the study of folk illusions also presents an excellent opportunity to address a frequently unmentioned aspect of folk knowledge: namely, traditionalized knowledge about and/or awareness of embodiment. It is nothing new to suggest that people are intimately aware of their own bodies, but more than the unique, subjective kind of knowledge that accompanies embodied consciousness, folk illusions represent a kind of folk knowledge that highlights cultural awareness of the sometimes misleading, illusory quality of perceptual processes as a shared aspect of individuals' subjective realities. To this point, the layperson's awareness of perceptual processes has been largely unrecognized or ignored by the scholarly disciplines that have considered perception and illusions most thoroughly. Instead, scholars of perception and illusions tend to argue the opposite—that the layperson, the everyday individual, the naive subject, and the average nonspecialist remain oblivious to the nature of illusory perceptions.

A few examples are enough to illuminate perception studies' seemingly obligatory references to the naivete of the folk. Psychologist Stanley Coren's chapter "Sensation and Perception," which appears in the first volume of the twelve-volume set *Handbook of Psychology* (2013), begins with a pointed assertion about the everyday individual's understanding of perception's relationship to reality: "Most people have a naïve, realistic faith in the ability of our senses to convey an accurate picture of the world to us. For the proverbial 'man on the street,' there is no perceptual problem. You open your eyes and the world is there. According to this viewpoint, we perceive things the way that we do because that is the way that they are. We see something as a triangular shape because it is triangular. We feel roughness through our sense of touch because the surface is rough" (101). The naive subject remains a central (and often unquestioned) aspect of the experimental sciences of mind. For the prototypical "psychologist in the lab," the idea of the naive "man on the street" represents the unphilosophical, unscientific minds of people who live their lives in some (usually unnamed) psychocultural space outside of scientific enlightenment. Coren happily corrects the proverbial man on the street's problematic position via a respectable scientific example, the Müller-Lyer illusion:

> Unfortunately, the man on the street [is] wrong, since perception is an act, and like all behavioral acts, it will have its limitations and can sometimes be in error. One need only look at the many varieties of visual-geometric illusions that introductory psychology textbooks delight in presenting to verify this, such as the Müller-Lyer illusion . . . , where the upper horizontal line with the wings out appears to be significantly longer than the lower horizontal line with the inwardly pointed wings despite the fact that a ruler will prove that both lines are physically equal in length. In such simple figures, you can see lines whose lengths or shapes are systematically distorted in consciousness due to the effects of other lines drawn in near proximity to them. Such distortions are not artifacts of art or drawing but reside within the mind of the observers. These "errors" are the basis of the correspondence problem, which simply asks the question, "If the senses are accurate recorders of the environment, why is the case that what we perceive can sometimes systematically differ from the physical reality?" (101–2)

We do not deny that people out in the world usually assume that their sensory perceptions accurately correspond to reality. How else do we manage walking around, pouring a cup of coffee, or kissing a lover? But there are many possible ways for humans (even laypersons) to assume the reliability of reality without remaining completely naive to illusory perceptions. We wonder how an individual's lack of awareness of the scientists'

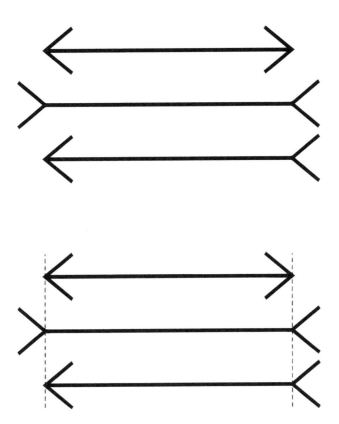

Fig 1.2. In the Müller-Lyer illusion, the line with inward-pointing arrows appears to be longer. (Fibonacci, CC BY-SA 3.0, https://commons.wikimedia.org/w/index.php?curid=1792612.)

optico-geometric creations, like the Müller-Lyer illusion, rules out awareness of illusory perceptions out in the world.

Psychologist Ken Nakayama's chapter "Modularity in Perception, Its Relation to Cognition and Knowledge" in the *Blackwell Handbook of Sensation and Perception* (2005) deploys proverbial wisdom in support of Nakayama's argument that laypeople do not question the accuracy of the senses: "Although, as lay people, we take it for granted that perception offers us sure knowledge of the world, perception is deficient in some rather fundamental ways. It doesn't reveal all that we know about the world. Unaided, it says very little, for example, about the microscopic structure of matter. Perception can also be mistaken at the scale of the everyday, showing a host of well-known errors, so-called illusions. Yet, these obvious limitations do not shake our confidence in perceptual experience. Seeing is believing and we take perception as a reliable source to bolster our most deeply held beliefs" (737). How does the child's disbelief in the rubbery transformation of his pencil align with such a simple version of the layperson's sight-to-belief tendencies? How much awareness of perception's illusory tendencies must the layperson demonstrate in order for the rational-minded scientist to dismiss charges of naivete? Recall that Pomerantz only published his scientifically parsed explanation of the Rubber Pencil illusion in 1983. Are we left to suppose that, before then, cultural notions of perception were not adequate for recognizing the illusory quality of the pencil's ephemeral rubbery wobbling?

Research scientist Jacques Ninio, in his insightful and useful *The Science of Illusions* ([1998] 2001), frames humans' ubiquitous naivete as its own, grand illusion:

> One illusion, possibly the strongest of all, is the one that makes us believe that we have a direct hold on reality. The work of interpretation conducted by perception never comes to light and leaves no other trace than its final result. Although the natural world that surrounds us is revealed to us by way of electrical signals exchanged among neurons, we have the impression of contact with it at a distance. The tennis player feels the ball at the end of his racket; the handyman feels the resistance of the screw at the tool's tip. In the laboratory von Békésy created tactile localization "at a distance" by transposing to the domain of touch the principle of stereo sound or vision. Subjects to whom one transmits two synchronized vibrations to the end of the index finger respectively are able, after practice, to localize the sensation at a point intermediate between the two fingers, even if the fingers are spread apart. In an experiment of the same kind, a subject who receives synchronized vibrations on his two thighs, just above the knees, manages, after practice, to localize a vibratory sensation between his two knees, although they are spread apart. (181)

At some level of explanatory exactitude—what Daniel Dennett calls the *subpersonal* level of explanation—it is undoubtedly true that phenomenological experience does not supply folk knowledge with insights of naturally unobservable phenomena. But why is Ninio so comfortable assigning a kind of naivete to the layperson's technologies (a screwdriver and tennis racket), only to assign a kind of insight to the scientist's vibrating laboratory devices? Why should we assume that the layperson does not (or cannot) recognize the tenuous philosophical grounds upon which she bases her faith in the swing of a tennis racket or the twist of a screwdriver? Does the tennis player not know best that her shots—though perceived as well struck—are wont to go awry for largely unobservable reasons?

You probably already guess our response. If many people have experienced or have performed folk illusions like the Rubber Pencil illusion by the time that they reach their teens, then the prototypical person on the street cannot be as naive to the nature of perceptual errors and perceptual illusions as scientific characterizations of the naive layperson would have us believe. To be clear, we do not deny the fact that unconscious perceptual processes affect the phenomenology of perception of laypeople (and nonlaypeople) in profound ways. We, obviously, do not deny the fact that experimentally derived data opens an otherwise closed window on the inner workings of such unconscious processes. As we mentioned, we lean heavily on that data in this book. We do, however, deny the notion that "the work of interpretation conducted by perception never comes to light and leaves no other trace than its final result" (Nino [1998] 2001, 181).

On the contrary, we find that the unconscious, interpretive processes of perception *do* leave traces. They are cultural traces captured in a kind of folklore, folk illusions. As we move through the artistic and philosophical features of folk illusions, we look to fill up the void of naivete. We look to unveil the folk's dynamic awareness of perceptual illusions.[5]

Settling Terms

The terms in our name for the genre, *folklore* and *illusions*, share important etymological complexities: (1) Both terms have proven themselves difficult for scholars to define, and (2) both terms are used in everyday language in ways that do not necessarily align with their respective scholarly definitions. Their meanings blur in the wash of many different methods and many different aims. Scientists and philosophers by nature prefer clear definitions that serve specialized purposes, and people—real people—concern themselves with the more important and just as noble task of getting on with their lives.

Illusion presents the older definitional problem. The English word *illusion* derives from the Latin *illudere*, meaning "to make sport, to jest or mock at, and to trick."[6] Several folk illusions are, in fact, performed in order to trick and make sport of someone (see, for example, our description of Snake in the Cooler below), and social notions of illusions' trickiness are captured in traditional phrases: "My mind is playing tricks on me" and "You must have a trick up your sleeve." Historically, a kind of distrust—not unlike the distrust that follows being tricked—of sensory and perceptual mechanisms permeates writings on myriad philosophical subjects such as nature, sensation, perception, physics, body, and mind. Ancient Greeks elevated reason above perception largely because of perception's susceptibility to illusions. Plato, and many after him, famously argued that reason exists outside of the realm of physical sensation and that the mind (i.e., the intellect) is characterized by our ability to think about the senses. For Plato, the fact that we misperceive is less important than the fact that we can recognize misperception as such.[7] Our name for the genre does not imply any misperception or breakdown of normal psychoperceptual processes; instead, illusions arise from the normal psychophysiological operations. We can safely say that pencils are not actually rubber during a performance of Rubber Pencil, just as we can say that the child who sees the pencil bend is not misperceiving. It is "normal," physiologically and socially, for any given performance of a folk illusion to succeed.

In western philosophical traditions since at least the time of Pyrrhonism, references to a common set of specious percepts like the bent oar in the water, trompe l'oeil in paintings, and the distortion of shapes in concave or convex mirrors have been repeatedly deployed as evidence against the accuracy of the senses.[8] Illusions' regularity, it seems, forces the skeptic's worth into the foreground of experience. If seeing is believing, doubting is knowing. But like one of Douglas Hofstadter's strange loops, doubt itself compels human minds along endless garden paths. Saint Augustine of Hippo, writing in the second book of his *Soliloquies* (ca. AD 386), phrased the dilemma in the contexts of that which we deem false: "We speak of a false tree which we see in a picture, a false face which is reflected in a mirror, the false motion of towers as seen by those sailing by, a false break in an oar in the water. These are false for no other reason than that they resemble the true" (trans. Paffenroth 2000, 66). Like the term *folklore,* the term *illusion* is often utilized in such a way as to suggest falsity, unreality, and untruth. We hear and say that false things are "just folklore" just as we hear and say that a false thing is "just an illusion."

Unlike folklore, however, illusions—or, more specifically, illusions' special relationship to "truth" and "actuality"—have been systematically deployed by scholars of mind to examine the true nature of reality vis-à-vis perceived reality. As Augustine suggests, illusions' proximity to perceptual reality presents philosophers and scientists with a fruitful, comparable juxtaposition. The only reason we think of the Rubber Pencil illusion as such is the illusory pencil's relationship to actual rubber rods. Maybe this is why the child wiggles her pencil up and down, to and fro, time after time, to perceive in one phenomenological moment the juxtaposition of the true and the false. Elena Pasquinelli, who focuses on the intersection of philosophy, cognitive science, and illusions, notes that illusions concentrate awareness in productive ways: "Illusions are perceptual phenomena that are at the same time systematic, robust, and surprising. Surprise puts the subject in contact with expectations that she did not necessarily hold in an explicit fashion, thus providing illusions with an epistemic value. At the same time, illusions have a heuristic value, in that they give us access, personally and as researchers, to perception and its characteristics" (2012, 73–74). Agreeing with Pasquinelli, we look to add the dimensions of social discourse and play. That is, personal epistemology and scholarly insight do not compose the whole of knowledge. Human knowledge is also created and maintained in systems of culture—especially cultural traditions.

Trying to walk this tightrope between intellectual values and empirical realities, folklorists have debated the exact meaning of *folklore* since William Thoms coined the term in 1846. Even though the definitional problem of *folklore* is younger than *illusion*'s, it is no less convoluted. In his introduction to Richard Dorson's *Handbook of American Folklore*, W. Edson Richmond begins with the declaration that the folk idea that "there are more definitions of folklore than there are folklorists" had long been a cliché (1986, xi). By that time, so many books and articles had regurgitated the fact that a *Funk and Wagnalls Standard Dictionary of Folklore, Mythology, and Legend*, published in the middle of the twentieth century, had listed twenty-one separate definitions for folklore.

The tradition continues into the twenty-first century. The American Folklore Society (n.d.) lists eleven definitions from eleven different authors on its web page, "What Is Folklore?" In that list, we find Henry Glassie's metacommentary on the definitional activities of the profession:

> "Folklore," though coined as recently as 1846, is the old word, the parental concept to the adjective "folk." Customarily folklorists refer to the host of published definitions, add their own, and then get on with their work, leaving the impression that definitions of folklore are as numberless as insects. But all the definitions bring into dynamic association the ideas of individual creativity and collective order.
>
> Folklore is traditional. Its center holds. Changes are slow and steady. Folklore is variable. The tradition remains wholly within the control of its

practitioners. It is theirs to remember, change, or forget. Answering the needs of the collective for continuity and of the individual for active participation, folklore . . . is that which is at once traditional and variable. (Glassie 1989, 24–31)[9]

For now, we will avoid the temptation to add our own definition of folklore to the list, but let us point out that folk illusions involve many agreed-upon, central characteristics of folklore proper:

- Folk illusions are usually passed along via interpersonal interaction.
- Folk illusions manifest in formulaic variants.
- Folk illusions accentuate philosophical, expressive, and artistic communicative skills.
- Folk illusions maintain formal and performative traditions.
- Folk illusions are performed in small groups.

Ultimately, general definitions for a cultural phenomenon require leeway for reinterpretation according to empirical circumstance.

Dell Hymes called for a kind of folkloristics that appeals to scholarly desires to explain "some fundamental aspect of reality" (1975b, 347). In communal traditions, people isolate intersubjective attention to and care for discrete aspects of reality. In as much as epistemologically concerned scholarly disciplines like experimental psychology, philosophy, or cognitive science lack the ability to capture contextualized cultural performances, we deploy folklore as a tool for examining everyday attention to the "realities" of perception.

Folkloristics, of course, is not the only discipline to use the term *folk*, and the construction of our phrase, *folk illusions*, parallels the construction of phrases that include the term *folk* from the more scientistic examinations of epistemology. Gregory Schrempp has exposed the differences between such terms as *folk science, folk physics, folk psychology,* and *folk classification* used frequently in cognitive science and *folk* as folklorists use the term. In the former, "'[f]olk' is usually used as a label for principles or systems of cognition, especially classification, that have developed outside of the formalized procedures characteristic of academic disciplines, especially those of mathematics and the physical sciences" (1996, 191). *Folk*, in the scientist's usage, often signifies a naive, untested system of thought. As we have said, we suggest no naivete in our use of *folk*; instead, we mean to highlight those aspects of traditionalized, expressive communication necessary for a successful performance of any given folk illusion. From our point of view, the child's performance of Rubber Pencil is important not because the child is naive to Pomerantz's scientifically founded translational and rotational equation, but because the child has learned to perform the illusion without needing the equation, without any interest in the equation.

Illusions and Culture

Folk illusions demonstrate the fact that people enjoy illusory perceptions. Especially, people enjoy elaborate illusions, which is to say people enjoy magic. We need not list the obvious contemporary examples of twenty-first-century Hollywood films that make billions of dollars via verisimilar visual presentations of fantastical and otherwise impossible feats, for the stage and street magician's art form is ancient. Beloved magician (and hated debunker) James Randi once quipped that magic is the world's second oldest profession (1992, xi). Professional magicians often mention Egyptian illusionist Dedi, who impressed King Cheops around 2600 BCE by cutting the heads off of domestic fowl and one calf only to return the decapitated creatures back to life as an important early staged performance. The paintings on the walls of a burial chamber in Beni Hasan, Egypt, painted around 2000 BCE, purportedly depict two men performing a cups-and-balls routine. Some, like Renaissance playwright Christopher Marlowe, have rather controversially suggested that Moses' turning of rods to snakes in the Book of Exodus presents another ancient example of staged magic.

Hieronymus Bosch's *The Conjurer* (ca. 1502) more clearly depicts a street magician's performance of the cups-and-balls routine in all of its culturally perceived suspicion. The main characters in *The Conjurer*'s foreground—the magician and his most interested spectator—face each other across a sheeted table while other slightly backgrounded audience members look on. Like the misdirectional techniques of performed magic, it is in the background characters of Bosch's painting that we plainly see the painter's legerdemain: a deviously placed boy with his hands in the most interested spectator's robes and a purposefully placed man with his hand on the most interested spectator's coin bag.

Somewhat surprisingly, the often dubiously characterized methods of magicians have received a good amount of recent (re)consideration in the contemporary sciences of mind. In the past decade, several cognitive, neuro-, and psychological scientists have keyed onstage magicians' deep understanding of their audiences' mental and perceptual processes—such as attention, awareness, and choice—as prompts for experimental inquiry (e.g., Kuhn, Amlani, and Rensink 2008; Cavina-Pratesi et al. 2011; Barnhart 2010).

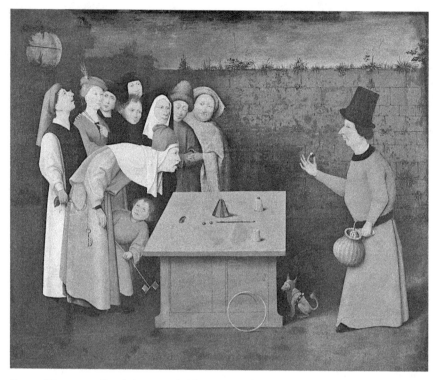

Fig. 1.3. Hieronymus Bosch (1450–1516), *The Conjurer*. (By Hieronymus Bosch and workshop—The Yorck Project: 10.000 Meisterwerke der Malerei. DVD-ROM, 2002. ISBN 3936122202. Distributed by DIRECTMEDIA Publishing GmbH. http://art-in-europe .info/2014/10/20/hieronymus-bosch-the-conjurer-1502-in-hd/, public domain, https://commons.wikimedia.org/w/index.php?curid=148032.)

In their excellent popular science book, *Sleights of Mind* (2010), neuroscientists Steven Macknik and Susana Martinez-Conde go so far as to suggest that understanding magic tricks will provide access to the core aspects of human experience: "Magic tricks work because humans have a hardwired process of attention and awareness that is hackable. By understanding how magicians hack our brains, we can better understand how the same cognitive tricks are at work in advertising strategy, business negotiations, and all varieties of interpersonal relations. When we understand how magic works in the mind of the spectator, we will have unveiled the neural bases of consciousness itself" (2010, 26). Like the ancient philosophers who precede them, Macknik and Martinez-Conde clearly believe that the pathway to an understanding of reality passes directly through human experiences of unreality.

Children and youth who perform folk illusions like Rubber Pencil, however, invest neither the time nor the resources correlative to professional conjuration. Moreover, we will be especially careful when considering the implications of folk illusions' relationship to cultural notions of magic because magic largely considered coincides with so many other aspects of culture. In truth, the amount of scholarly *and* cultural attention devoted to magic in supernatural subjects like religion, belief, and ritual or to magical beings like spirits, witches, and fairies is probably impossible to catalog or summarize with fidelity.[10] For our purposes, we find that Alan Dundes productively narrows the definition by focusing on the acknowledged causes and effects of a given phenomenon: "Magic . . . involves some kind of instrumental or causal process or procedure by means of which events are produced or controlled. Many writers on magic have specifically drawn attention to the 'coercive' nature of magic. Leaving aside the thorny question of the rationality of magic—though conceivably all magical acts conform to one or more culturally relative logics—it can be legitimately argued that magic in the more rigorous, narrow sense of the analytic term does imply the influencing or manipulating of nature in some causal way" (Winkelman et al. 1982, 46–47). Certainly, folk illusions apply a causal procedure to the body of the experiencer in order to influence a particular illusion, and given Dundes's definition, we might think of folk illusions as *body magic*. But as we will see, even this general application of the notion of magic to folk illusions falls apart, for example, when nonmagical, unexplained causal mechanisms (like pseudoscience) are believed to induce the intended phenomenal experience.[11]

Regardless, we cannot deny that magic and perceptual illusions are frequently connected in cultural expression. We find many examples of special attention to perceptual illusions, for example, in the narrative genres of folklore. Stith Thompson's *Motif-Index of Folk Literature* ([1958] 2001) lists over one hundred examples and variants of folk narratives featuring episodes of "Deception by Illusion" (K1870), found in dozens of cross-cultural narrative traditions and several areally distinct language families.[12] In Iceland, we find legends of ships that are made to look like islands via the illusion of reeds in order to trick an enemy's fool-hearted advance (K1872.2). Stories of sham blood that is used to create visual and somatic illusions of bleeding and death are told in France, Japan, Indonesia, and the West Indies (K1875). We find examples of popular legends about luminescent visions, like the will o' the wisp in Finnish-Swedish, Scottish, Icelandic,

British, Indian, Dutch, and African traditions (F491). In the Irish legend of King Dathi, a sponge lit on fire and placed into a dead king's mouth creates the illusion that the Dathi is breathing (K1885.1). Traditional stories involving the illusory answer of an echo are found in Greece, China, and Spain (K1887.1). Even the fantastic and anthropomorphized creatures of folk narratives succumb to illusions; Spanish and Japanese stories tell of dragons that are tricked into attacking their reflections in the mirror (K1052).

Religious, or sacred, beliefs in several cultures deal directly with illusions. The etiological functions of sacred beliefs often encompass the existential, and thus phenomenological, realities of being, and it is true that anomalous phenomenological experiences are sometimes categorized as spiritual—that is, having a metaphysical orientation. Hindu beliefs about the illusory qualities of perception, which rest at the core of the traditions' metaphysical stages of consciousness, provide an excellent example. Comparative mythologist Wendy Doniger argues that a fundamental search for a deeper understanding of illusion and of illusions' role in the first stage of consciousness permeates Hindu thought beyond the boundaries of intellectual, philosophic inquiry:

> Unlike other topics that only erudite Indian philosophers wrestled with, illusion got into the very fabric of Hindu culture, so that just about everyone knows about *maya* and the difficulty of telling a snake from a rope. *Maya* (from the verb *ma* ["to make"]) is what is made, artificial, constructed, something that seems to be there but has no substance; it is the path of rebirth, the worship of gods with qualities (*sa-guna*). It is magic, cosmic sleight of hand. *Maya* begin in the earliest text, the *Rig Veda* (1.32) in which the god *Indra* (the first great magician; magic is called Indra's Net [*indra-jala*]) uses his magic against his equally magical enemy Vritra (for all the antigods are magicians): Indra magically turns himself into the hair of one of the horses' tails, and Vritra magically conjures up a storm. Magic illusions of various sorts play a crucial role in the Valmiki *Ramayana*, in the shadow of Sita of later traditions, and in the Hindu thinking across the board. (2009, 516)

Anyone with even a mild case of ophidiophobia (the fear of snakes) who has happened upon a darkened rope curled under a canoe or an unexpected garden hose lying across a path knows the jolt of fear that accompanies the illusory perception of a snake.

Experimental studies offer scientific confirmation of what ancient narratives, like the *Upanishads*, understood centuries ago: The illusory perception of snakes is primal. According to recent neurological tests of the Snake Detection Theory, humans and other primates visually detect snakes much faster than other harmless objects and creatures because certain neurons "respond selectively to snakes and in ways that facilitate their rapid visual detection" (Van Le et al. 2013, 19003). Similar research with patients who have a pronounced fear of spiders (arachnophobia) suggests that while highly phobic individuals are not better at identifying actual spiders, they are more likely to assume that anything looking even remotely like a spider *is* a spider (Becker and Rinck 2004).

A Serpentine Folk Illusion in Pensacola, Florida

K. Brandon Barker

Having families that prefer to take summer vacation on the white, sandy beaches of Florida's Gulf Coast, my wife and I spend about a week or two every summer somewhere in the Panhandle. In 2012, we stayed in a comfortable, humble neighborhood just across the road from Pensacola Beach. There, visiting families mixed together in a friendly, if short-lived, community while swimming in the neighborhood's pool and playing on the neighborhood's enclosed streets. One family, who strangely seemed to be the misfit, lived permanently in the neighborhood—just across the street from the home we were renting. Their bluish gray, low-country home stood on stilts. Within the stilts, in the shade of the home, the owners frequently played music and greeted walkers heading out to and back from the beach. On one such occasion, the owner, Tom, asked us if we would like a drink. We, respectfully, agreed, and Tom handed my wife a beer. Nodding off to his left, Tom said, "There's plenty more in the cooler."

I walked over to the cooler and eagerly lifted its lid. Snake! That old, primal fear shot up my spine as I retreated. Tom and my wife laughed aloud. To my relief, Tom hurried over and lifted the lid to show me the (exceedingly lifelike) rubber snake whose head he had attached to the bottom of his cooler's lid via fishing line. An excellent, if torturous folk illusion!

But is the Snake in the Cooler really an illusion? It remains a difficult question to answer. Certainly, my initial perception and reaction were not incorrect. Had the snake actually been real, my rather hasty retreat might have saved my life. In the context of perception, the snake was real—that is, the rubber snake possessed all of the visually perceptible physical characteristics of a real snake. In fact, the rubber snake's attachment to the lid also meant that lifting the lid forced the rubber snake to lift up as if it were lunging for me. Surely, we cannot expect the perceptual systems of humans to compete with technologically assisted mechanical representations of things like snakes, which we are experientially, culturally, and possibly biologically made to fear.

We can say with more certainty, though, that the Snake in the Cooler is a living folk tradition among the locals of Pensacola Beach. Two days after my experience with Tom's cooler, a few of us headed out in the dark hours of the morning in order to fish on the last day of red snapper season. The trip was a success, and we returned in the early afternoon with our limit of snappers and a gathering of trigger fish and trout. Arriving at the dock, our charter's wily captain, Captain Chuck, kindly offered to clean and bag our catch. Just as Captain Chuck started with the first snapper, he asked me to walk down the pier and to bring him a bag of ice from the red cooler, the red cooler that had clearly been resting in the same place all morning, the red cooler that had a barely visible fishing line attached to the inner side of the lid. This time, I suffered no illusion.

Rope snakes and illusory spiders are good examples of the many illusions that readily occur as a part of people's everyday comings and goings. Yes, illusions and references to illusions fill up multiple types of cultural discourse because illusions force humans to deal with skewed phenomenological experience, but it is also true that humans are interested in illusions because we experience them and know them as a part of our lives. As we mentioned above, several illusions that occur naturally—that is, occur readily in a situation frequently present in nature and do not rely on humanly constructed environments or technologies—were reported by classical authors. There is no reason, however, to assume that these authors were the discoverers of these illusions. It makes much more sense to consider these illusions a form of traditionalized, folk knowledge that was adapted by ancient thinkers to contribute to their philosophical systems.

The Waterfall illusion must have been known by many people. As Aristotle describes it, "When persons turn away from looking at objects in motion, e.g., rivers, and especially those which flow very rapidly, they find

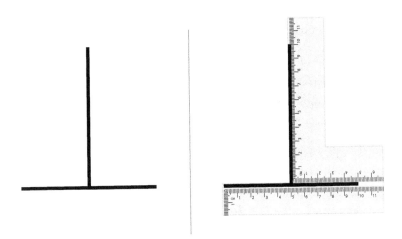

Fig. 1.4. The vertical line appears longer than the horizontal. (By S-kay, public domain, https://commons.wikimedia.org/w/index.php?curid=8845534.)

that the visual stimulations still present themselves, for the things really at rest are seen moving" (quoted in Gallop 1990, 89). As an experience of common environmental contexts, this illusion occurs in nature, and the same can be said for the Moon illusion, which involves seeing the moon as much larger on the horizon than at its zenith. Aristotle's mention is the first recorded, but the Moon illusion appears frequently in the history of philosophy, and even more frequently in people's lived experiences.

For philosophers and scientists, the facts that illusions appear in numerous oral traditions or that illusions are purposefully created during certain folkloric performances seem to have been less interesting than questions surrounding the identifiable roles that cultural influences may or may not play in the perception of any given illusion. Historically speaking, the Snake Detection Theory's argument for the innate quality of rope or snake illusions merely presents a recent addition to long-standing arguments about the biological versus cultural causes of illusory experience. And while illusions represent only a small part of the larger theoretical arguments concerning nature versus nurture, it is worth noting that culture took center stage in several important studies of perception and illusion in the context of optico-geometric illusions at the beginning and again in the middle of the twentieth century.

Polymath W. H. R. Rivers's "Observation on the Senses of the Todas" (1905) first examined cross-cultural susceptibility to certain optico-geometric illusions. Working with the Toda, a small pastoral community in southern India, Rivers tested several aspects of perception, including color discrimination, visual acuity, tactile discrimination, taste, and visual illusions. In those tests, Rivers noted an unexplained tendency for the Toda to show less susceptibility to the Müller-Lyer illusion than Europeans while showing more susceptibility to the Horizontal-vertical illusion (fig. 1.4). In the case of the two cultures' perception of the Müller-Lyer, Rivers offered the short explanation that the *savage* man "is less influenced by the figure as a whole" than the *civilized*.

A number of follow-up studies confirmed the cultural difference noted in Rivers's work with the Toda, and six decades later Marshall Segall, Donald Campbell, and Melville Herskovits explained the difference via their Carpentered World Theory:

> An example of a cultural factor which seems relevant is the prevalence of rect-angularity in the visual environment, a factor which seems to be related to the tendency to interpret acute and obtuse angles on a two-dimensional surface as representative of rectangular objects in three-dimensional space. This infer-ence habit is much more valid in highly carpentered, urban, European envi-ronments, and could enhance, or even produce, the Müller-Lyer and Sander Parallelogram illusions. This interpretation is consistent with traditional explanation of these illusions. Less clearly, the Horizontal-vertical illusion can perhaps be understood as the result of an inference habit of interpreting verti-cal lines as extensions away from one in the horizontal plane. Such an infer-ence habit would have more validity for those living in open, flat terrain than in rain forests or canyons. (1963, 770–71)

Like many issues in the scientific study of illusions, the validity of the Carpentered World Theory remains an ongoing debate. In current discus-sions of the Müller-Lyer illusion, for example, opponents of the Carpen-tered World Theory cite evidence that the illusion diminishes as detailed orientation profiles of the two-dimensional presentation of the illusion are closely examined or that current artificial intelligence models suggest that no "enculturation" is necessary for experiencing the illusion. Proponents still hold that cuboidal shapes experienced as a result of a carpentered cultural setting do affect perception of the illusion. However, they clarify that the smaller angles (correlative to the line that is perceived as shorter) must be presented between 90 degrees and 180 degrees in order for the two-dimensional figure to be adequately similar to the visual perception of an actual cuboid (Deręgowski 2013).[13]

While we fully expect that many cultures share genres of folklore akin to folk illusions, it is important to note that the examples covered in this book are not fully cross-cultural, so our project cannot necessarily support one side of the Carpentered World Theory debate or the other. Our work with folk illusions does stem, however, from a more global theoretical position that humans and human embodiment are best thought of as always already encultured. Undeniably, genetic and biological predispositions matter a great deal in questions of embodied experience (including perception), but conception (almost always) and gestation (always) occur in the extensively cultured bodies of our mothers. A newborn, we know, responds to cultural stimuli, like certain phonetic aspects of his or her mother's language.[14] Rituals and material customs like circumcision and clothing traditions inundate infants' bodies from the outset. There exists no precultural human embodiment, so we are skeptical of any nature-or-nurture dichotomy that requires a priori simplification of the constant intermingling of innateness and experience.

A (Rubber) Pencil in Plain Sight: Illusions, Childhood, and Triviality

Henry Petroski, a professor of civil engineering, begins his compelling book *The Pencil* (1992), on the technological development of the pencil, with an anecdote about Henry David Thoreau's packing lists for his famous trips into the woods of New England. Thoreau was a meticulous writer, thinker, and camper. Petroski mentions several items on one of Thoreau's packing lists, which the transcendentalist had prepared for a twelve-day trip into the woods: a tent, strong cord, old newspapers for cleaning, foodstuffs, matches, soap (two pieces). The list goes on. Peculiarly, Thoreau makes no mention of a writer's most important tool—his pencil. The apparent omission is amplified when we consider the fact that Thoreau had worked with his father, John Thoreau, a famous pencil maker.

Petroski offers an insightful possible explanation: "Perhaps the very object with which he may have been drafting his list was too close to him, too familiar a part of his own everyday outfit, too integral a part of his livelihood, too common a thing for him to think to mention" (1992, 4). It is a good question: How often do we notice the truly mundane? It is well known that Thoreau escaped to Maine's woods in order to break from the everyday, in order to test his personal resolve, and to shift his social perspective, but what do we make of the fact that Thoreau presumably unconsciously omitted his pencils from a list that he consciously scribed for the sole purpose of facilitating his break from the ordinary?

The modern pencil likely evolved from the stylus, which was used in the earliest forms of writing to scratch symbols into papyrus and wax. Etymologically, the name *pencil* comes from the Latin term for a kind of small paintbrush, the *pencillu*. Pencils filled with graphite and possessing an attached piece of rubber for erasing took shape in the middle of the sixteenth century. Pencil makers in Europe, Asia, and North America made important, standard improvements to wood quality, point preservation, and aesthetic varnishes in the nineteenth century, and by the second decade of the twentieth century, America was producing upward of 750 million pencils a year (Petroski 1992, 205). A 2010 article in *The Economist* ("The Future of the Pencil," September 16, 2010) estimated the worldwide, annual production of pencils in the twenty-first century to be between fifteen and twenty billion.

Just as the pencil is a tool for adult business and art, the pencil is also a tool of the student. Along with the student's books, paper, clips, and backpacks, the pencil fades into the academic lives of little learners. Adults easily forget that the child—when deprived of her playthings and pressed to invent her own toys—possesses the undeniable ability to reshape the intended purposes of many tools for the sake of play. Children's folklore is full of ways to make the instruments of learning into instruments of play. Ruled paper becomes wadded projectiles, airplanes, and ritualesque mediums for divination. Paper clips become sculptures, poppers, and pokers. Rubber bands become bracelets, slingshots, and bouncing balls. And the pencil—rigid and versatile—becomes sporting equipment for the folk game, pencil break, which involves two players taking turns striking each other's pencil until one player's pencil breaks into halves. Pencils can be loaded with wetted bits of paper—spit wads—and used as a catapult. A sharpened pencil when thrown at the otherwise unadorned, dropped ceiling tiles that populate so many modern school buildings will stick like a dart until some janitor has time to retrieve a ladder. The pencil can magically be made rubber.

We know that by the turn of the twentieth century, most schoolchildren in America had replaced their chalk and slate boards with pencils and paper, but like many of the folk illusions we discuss in this book, we cannot be certain as to how long children have been performing the Rubber Pencil illusion. Children have always had access to many different long, rigid objects—namely sticks—that make a performance of such an illusion possible. Nothing has changed in that regard; in 2014 while observing a group of four- to seven-year-old boys and girls, we saw performances of successful Rubber Pencil illusions with long Lego pieces!

To finish getting us started, then, we must propose one last introductory question: Can children's play with mundane objects like school pencils during unimportant daily routines like eighth-grade science (or—even less important—recess) really provide answers to clearly important problems like cultural influences on human embodiment or the experiential basis of reality? It is a question with many layers. Is childhood inherently trivial? Is play? What about folklore, or illusions? The answers to these layers of questions, we believe, must contradict the questions' implicatures, for what we come to find—when we take seriously children's traditionalized play with embodied illusions—is that triviality itself is seldom trivial. Impetuous labeling of "triviality" is rarely unproblematic, and whether or not something is actually trivial is a question that can only be answered via deep, systematic analysis in the contexts of specific social situations. Unfortunately, much that gets marked as "trivial" never undergoes such specific analysis.

In her "Early History of Perceptual Illusions" (1971), Dorothea Johannsen wonders—more than ten years before the publication of Pomerantz's Rubber Pencil study—why so many illusions that she catalogs were not noticed or were not described in print until relatively late. She speculates, "One reason may be the general insensitivity of man to those items in his environment that are not immediately important to him" (1971, 138). Considering the well-known counterexample of rectilinear illusions famous in examples of Greek architecture, Johannsen observes that a builder who wishes to create a pleasing effect in a temple would look critically at the details and their contribution to the total structure. The Parthenon's architects might have noticed the effect of using true horizontals and verticals in particular locations and contrived ways to compensate. In Greek architecture, Johannsen identifies a context of cultural importance—a context where illusion matters.

One year before Johannsen's historical study, the psychologist, folklorist, and play theorist Brian Sutton-Smith put forward his elegant and powerful argument that the folklore of children—such as children's traditional jokes, rhymes, games, and toys—had been largely ignored by psychologists because the subject seemed too unimportant for serious study: "Probably most academic psychologists of the recent era would feel that there was something slightly worthless in studying such a subject matter, being confined by their self-respect to 'more important' matters such as eating, sex, and work. As we can define childlore partly in terms of its 'triviality,' it follows also that most serious persons will find it too trivial to study, that at this historical time there will be a 'triviality barrier' against its serious pursuit" (1970, 4–5). We have already mentioned several convergences of interdisciplinary concern—embodiment, illusion, development, culture—that shape our thinking on folk illusions, so triviality is another important convergence.

In their influential book on conceptual integration, or blending theory—a theory discussed in further detail in chapter 2—Gilles Fauconnier and Mark Turner articulate a commonly held proposition for contemporary sciences of the mind: "Cognitive scientists have shown that many feats that we find easy—categorization, memory, framing, recursion, analogy, metaphor, even vision and hearing—are exceptionally resistant to scientific analysis. They turn out to be the things that are hardest to explain. Syntax before Harris and Chomsky, framing before Bateson and Goffman, and analogy before Gentner, Hofstadter, and Holyoak were surely recognized as pervasive, but the need for their systematic study was not perceived. In a sense it could not be perceived, because there was no framework or set of techniques within which to ask questions systematically" ([2002] 2008, 59). Making the same point, folklorist Elliott Oring argues that folklore presents another example of pervasive trivia: "From trivia, we are directed to matters of gravity and consequence" (2012, 321). Oring takes as his proof the fact that folklore—to those who know it and perform it—remains interesting. He adds, "If folklore is interesting, this miscellany, in some manner, must disturb presumptions about the way the world appears and is understood" (320). Yes! Of all the points we will make, we can be most certain that youths' interest in folk illusions testifies to their desire to disturb their perceptual processes.

Considered together, Johannsen's and Sutton-Smith's arguments suggest that children's folkloric play with illusions constitutes a doubly trivial aspect of culture. Given this, we cannot be surprised to find that so many folk illusions we catalog and analyze in this book have gone unnoticed. Like Thoreau's omission of the pencil, how difficult is it for the seeing adult to think back on Rubber Pencil as a formative interaction? How difficult is it for the scientist drawing some new version of an optico-geometric illusion to focus his attention on the cultural technologies of pencil and paper that make the illusion possible? How difficult is it for the teacher, tasked with conveying the vastness of the Milky Way to a group of twelve-year-old boys, to allow playful performances of Rubber Pencil to interrupt her lessons? How easy has it been to leave folk illusions off of our lists?

Notes

1. While Pomerantz hypothesizes that the illusion results from "visual persistence" (i.e., afterimage) that occurs very early in the visual processes, a more recent study of Rubber Pencil by Lore Thaler et al., "Illusory Bending of a Rigidly Moving Line Segment" (2007), argues that such afterimage densities, which occur early in visual processing, cannot—by themselves—explain the experience of the illusion and that more active processes of perceiving spatiotemporal displacement are involved. We take up the notion of active perception in chap. 4.

2. Commonly associated with Piaget's developmental stages, questions around the child's understanding of object permanence have prompted much experimental study in, for example, visual attention and searching tests. For Piaget, the child must understand the connections between visual and tactual experience before understanding categories of object characteristics (e.g., size and shape, rigid or elastic, animate or inanimate). One such recent study, Schaub, Bertin, and Cacchione's "Infants' Individuation of Rigid and Plastic Objects Based on Shape" (2013) argues that twelve-month-old children showed diagnostic capabilities for object individuation based upon experience with the shape dynamics of rigid as compared to malleable plastic objects. After playing with a rigid or malleable object, children were given the opportunity to search for their object in an empty box. Children who played with a rigid object searched longer for a second object if (by way of an unobserved switch) the object in the box did not match the original shape and size of the rigid object. Other studies question Piaget's claim that infants who do not yet have the ability to grasp objects do not understand object properties. Using visual preference tests, for example, Bower (1966, 1971) argued that infants as young as four weeks old recognize and show visual attention to novel objects.

3. In his short essay "Seeing Is Believing" (1972), Dundes alludes to a number of colloquial and traditional American English phrases that articulate the primacy of sight. In light of these, he argues that Americans would be well served to become more aware of their seeing-is-believing cognitive biases. We might only add the visually (and haptically) grounded folk phrase *there are two sides to every coin*.

4. Our understanding of embodiment has benefited from Maclachlan's fine study, *Embodiment: Clinical, Critical, and Cultural Perspectives on Health and Illness* (2004). Therein, he provides a much fuller discussion of the term *embodiment*'s definition and its development (2–22).

5. Our argument—that people analyze and complicate their perceived world as a part of cultural traditions—falls out from a broader, general premise of folkloristics: communities,

by and large, produce critical traditions of aesthetic appreciation and epistemological inquiry that can and often do function in ways comparable to and in competition with high art, academic philosophy, and experimental science (Glassie 1989; D. Goldstein 2007).

6. *Oxford English Dictionary Online*, s.v. "illusion," accessed July 28, 2015, http://www .oed.com/view/Entry/91565.

7. For a strong description of Plato's position, see Cooper's "Plato on Sense-Perception and Knowledge" (1970).

8. For a short history, see Lloyd, "Observational Error in Later Greek Science," in *Methods and Problems in Greek Science: Selected Papers* (1991).

9. Quoted from American Folklore Society (n.d.).

10. Often associated with nineteenth-century theories of cultural evolution, the study of magic's relationship to the other major epistemological forces of culture (i.e., religion and science) can be identified in canonical works like Tylor's *Primitive Culture* (1871), Frazer's *Golden Bough* ([1890] 1915), and Mauss's *General Theory of Magic* ([1950] 1972). O'Connor's encyclopedia essay, "Magic" (2010), provides a fine synopsis and short bibliography of the folkloristic study of magic, which has highlighted the fact that magic and magical customs *do not* or *have not* disappeared with the arrival of scientific enlightenment. Davies's small book, *Magic: A Very Short Introduction* (2012) also provides a useful—more interdisciplinary— introduction to the study of magic. The *Current Anthropology* article "Magic: A Theoretical Reassessment [and Comments and Replies]" (Winkelman et al. 1982), presents an interesting interdisciplinary snapshot of magic studies in the second half of the twentieth century. (We borrow Dundes's definition of magic from his reply to Winkelman.) Simon During recently articulated a compelling argument in his *Modern Enchantments: The Cultural Power of Secular Magic* (2002) that magic (i.e., conjuration) with its special relationship to cultural triviality, technological innovation (e.g., automata and early cameras), and spirituality helped to usher in modernity.

11. This is the case for some variants of levitation illusions, like Lift from a Chair (see the catalog entries F3, F3a, and F3b in the appendix).

12. Despite enduring a number of critiques such as problematic overlapping and Eurocentrism (see Dundes 1997), Stith Thompson's *Motif-Index of Folk-Literature* (1958) remains an important folkloristic endeavor in the contexts of empirically grounded cross-cultural studies of narrative. Considering the recent attention given to empirical work on so-called narrative universals in the scientific study of emotion and cognition (see, for example, Hogan 2003), an interdisciplinary re-interpretation of Thompson's work feels appropriate.

13. Ninio explains his critique of the Carpentered World Theory in an excellent review article, "Geometric Illusions Are Not Always Where You Think They Are" (2014, 14). The gist of his critique is this: because the Müller-Lyer illusion is consistently perceived as an illusion even when the orientation of the optico-geometric image does not "align" with the orientation of real-world analogs to the two-dimensional image, then experiential precedents in the real world cannot explain the totality of illusory perceptions of the optico-geometric images. See also Zeman, Obst, and Brooks's study of the Müller-Lyer illusion when viewed by artificial neural networks; there, the scientists explain that artificial image-viewing systems do not need to be exposed to real-world analogs to the two-dimensional images in order to misjudge the length of the two lines of the illusion (2014).

14. Eimas et al. (1971) first demonstrated that infants at one to four months can decipher the phonemes of their mother tongue.

2

FOUR FORMS OF FOLK ILLUSIONS

IN THIS CHAPTER, WE INTRODUCE FOUR FORMS OF folk illusions. In so doing, we argue for the existence of the genre by highlighting its salient features—especially participant roles, morphological similarities, performance positions, priming periods, and ludic qualities.[1] Also, the four forms—Floating Arms, Twisted Hands, the Chills, and Light as a Feather—are representative examples of a verbal continuum that exists within the genre. Different folk illusions fall on this continuum according to the relative process-based, lyrical, or narrative quality of their respective verbal components. Variable positions on the continuum connote a range of aesthetic and communicative properties that are both traditionalized and necessary for creating the applicable embodied illusion.

Floating Arms

We begin with a folk illusion we call Floating Arms. To date, we have observed three variants. We documented one variant during our observations at St. Cecilia Middle School in Broussard, Louisiana, in the spring of 2011 (see D8a–D8c in the catalog in the appendix for other variants). During a performance of this variant, Child 1 instructs Child 2 to place her arms by her side so that the tips of Child 2's fingers are pointed toward the ground. Child 1 places her hands firmly on the outside of Child 2's hands. We refer to this position of Child 1 and Child 2 as the Floating Arms *performance position* (see fig. 2.1)—a performance position is the traditionalized position of all performers' bodies correlative to the kinesthetic action of any given folk illusion. Child 1 then instructs Child 2 to attempt to lift her arms up from her sides while Child 1 forcefully holds the other's arms down. Child 2 attempts to lift her arms against Child 1's applied pressure for approximately thirty seconds; we refer to these thirty seconds as the *priming period*.[2] At the end of the priming period, Child 1 releases her hold on Child 2's arms.

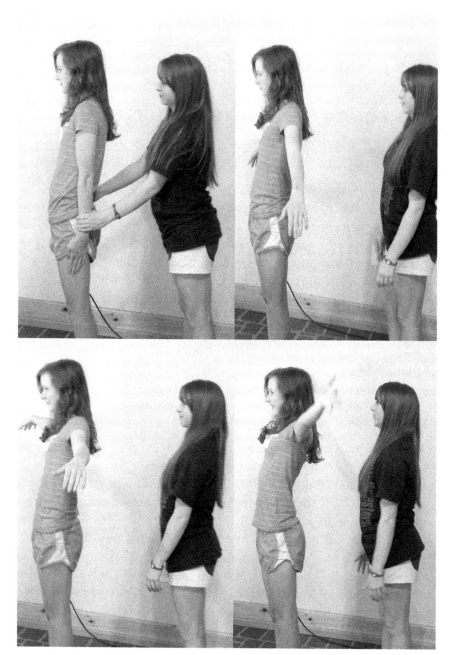

Fig. 2.1. In Floating Arms's performance position, the director holds the actor's arms at her side for approximately thirty seconds.

Child 2 then relaxes, slowly allowing her arms to rise from her sides. Child 2's arms feel as though they are floating.

While we find very little folkloristic study of Floating Arms, popular awareness of the form is certain. An assortment of websites dedicated to magic, the paranormal, and children's scientific activities refer to this illusion (e.g., Kidzworld, n.d.; Computer Science for Fun, n.d.). In our fieldwork, we find that most children and adults have heard of or played some variant of Floating Arms. We have witnessed performances of Floating Arms in Louisiana and Indiana, and we have gathered remembrances of performances from thirteen US states, as well as a remembrance from the state of Andhra Pradesh, India.[3]

In 1915, German neurologist Oskar Kohnstamm became the first scientist to recognize Floating Arms' intended illusion. Six years later, A. Schwartz and P. Meyer (1921) noted that the activity was frequently performed for fun by college students in England, and five years after that, Jayme R. Pereira (1925) reported that the activity was also performed by children in Brazil. For most of the twentieth century, neurologists and psychologists have attributed the illusion—commonly referred to as the Kohnstamm effect— to muscular and/or neural aftereffects following voluntary muscle contraction in a sustained direction.[4] Illusions resulting from muscle contraction aftereffects appear in other folk illusions, like Magnetic Rocks (see D2 in the catalog and our description of Magnetic Chicken Bones in this chapter), Falling through the Floor (see G1), and Legs through the Floor (see G3).

Neurologists, psychologists, and cognitive scientists have treated subjects with voluntary force, mechanical vibrations, and electronic pulses that induce muscle contractions in order to create aftereffect illusions at several different places on subjects' bodies—including the hip, the elbow, the neck, and the fingers.[5] Recently, scientists have suggested that such illusory aftereffects result "from the activity of a central network in which motor and somatosensory areas and the cerebellum all play key roles" (Duclos et al. 2007, 781), but it seems no experimental studies have seriously considered particular social or folkloric contexts that creatively categorize the experience of the Kohnstamm effect.

We know a few other variants of Floating Arms. We describe a few here, and we discuss others at length in chapter 5. In the first, which we observed a twelve-year-old boy from South Louisiana perform, the child places his hands in his pockets and presses outward for about thirty seconds. Interestingly, we have not gathered this variant from any girls. In the third variant, often referred to as the "door-frame illusion," children press the outside of their arms and hands against the inside of a door frame for an appropriate priming period. Different types of Floating Arms also vary in verbal descriptions of the illusion. In the spring of 2010, a twelve-year-old girl observed by one of our students in Lafayette, Louisiana, reported that "her arms felt weightless, as if they weren't there." While a sensation of weightlessness describes the illusion just as well as a sensation of "floating," we recognize that "weightlessness" and "floating," though similar, are not the same. Whether the arms float or disappear depends directly on the cultural discourse surrounding the performance.

In a recent issue of *Scientific American*, neuroscientist Vilayanur S. Ramachandran and researcher Diane Rogers-Ramachandran listed the door-frame variant of Floating Arms as a way to "get your hands to 'float.'" Not surprisingly, their son had reminded them of the variant:

> You can even get your hands to "float"—a well-known trick, sometimes called the Kohnstamm effect, was reintroduced to us by our son, Jayakrishnan Ram-achandran. Stand in the middle of an open doorway and use your arms to apply outward pressure on the two sides as if you were pushing them away from your body. After about 40 seconds, suddenly let go and relax, stand normally and just let your arms hang by your sides. If you are like most of us, your arms will involuntarily rise up as if pulled by two invisible helium balloons. The reason? When you apply continuous outward force, your brain gets used to this as the "neutral state"—so that when the pressure suddenly disappears, your arms drift outward. (2008, 63)

As folklorists, we wonder where Jayakrishnan Ramachandran learned the trick. We also recognize the varying description of the illusion; is it the hands or the arms that float? Are the arms floating or "being lifted by helium balloons"? Are they being lifted by helium balloons, or are they weightless? These culturally emergent descriptions are creative answers to philosophical questions about experience, and they represent an important aspect of the nature of the illusions themselves—namely, cultural awareness of the perceptual phenomenon.

That fact, as we explained in the previous chapter, has largely been ignored by the scientific community. A very recent study of the illusion published by cognitive scientists and neurologists in London measured subjects' responses to stopping, or inhibiting, the floating sensation in an effort to gauge whether or not motor inhibition is a part of the same neural processes that give rise to positive motor commands (Ghosh, Rothwell, and Haggard 2014), but the article makes no mention of the folkloric quality of the illusion *or* of whether or not the thirty-nine subjects between the ages of eighteen and sixty had performed Floating Arms prior to the experiment. In *The Body Has a Mind of Its Own* (2007), popular science writers Sandra and Matthew Blakeslee refer to the door-frame variant among other descriptions of "Illusions of the Body." After explaining several illusions that involve participants closing their eyes and taping mechanical buzzers to different parts of their bodies, the Blakeslees provide this description of Floating Arms: "Other illusions don't require you to close your eyes, such as the doorframe illusion children around the world enjoy at slumber parties. Stand in a door-way and press the backs of your wrists outward against the frame as hard as you dare for thirty seconds or so. Then relax your arms, step forward, and it feels as if your arms are being levitated" (2007, 35). While we know of no empirical research that suggests Floating Arms is a universally performed folk illusion, the Blakeslees' "door-frame illusion" morphologically coincides with everything we have observed. They even provide a new description of the aftereffect, "levitation." The Blakeslees go on to adumbrate the following value judgment of the door-frame variant: "This is child's play, a super-low-tech way of fatiguing your proprioceptors" (2007, 35), but the Blakeslees are not folklorists, and a scientifically biased triviality barrier keeps them from recognizing the discursive performance dynamics necessary for traditionalizing the illusion.

Magnetic Chicken Bones in Lafayette, Louisiana?

Claiborne Rice

Early in the process of our research on folk illusions, we realized that it would be nearly impossible to chance upon events of these activities being played, so we relied heavily on remembrances and reenactments. Occasionally, though, we found that we could initiate the folk illusions ourselves and observe people's reactions. We were at a party one evening discussing folk illusions when my son Will, who was between eleven and twelve years old, wandered by. Brandon and I had just been talking about how we had both been raised in the foothills of the Appalachian Mountains (eastern Tennessee and northeast Georgia, respectively), and we had each learned Magnetic Rocks (D2) as kids. Perhaps there was a geographic dimension to the distribution of illusions and variants? We could try it on Will to see if he knows the activity, I said, since he has grown up in Louisiana. We were inside, though, with no rocks to hand, but Brandon was inspired. We had been eating fried chicken among other things at the pot-luck.

"Hey," he called to Will, "I bet I can make chicken bones magnetic."

"No way," Will responded. "That's not possible."

"Sure," said Brandon with a straight face. "There are some bones on that plate right there; grab two and I'll show you how it's done."

Will got two bones, maybe a thigh and a leg, and came over.

"Hold them tight together, and keep holding them till I say stop."

Will took one bone in each hand, placed them end to end, and started pressing them together.

"Press hard," said Brandon. "Are you pressing as hard as you can?"

Will pursed his lips as he increased his efforts.

Brandon quickly grabbed a bone off the table and began passing it over and around Will's hands. "Oma-wa Kayuk ta-o," Brandon started, chanting some magical-sounding nonsense in a serious tone, interspersing with an occasional statement in English to the effect that "these chicken bones will become magnets, defying the science of the world."

Eventually Brandon told Will, "When I count down to one, you will try to pull your hands apart and discover that these bones are truly magnetic. 10 . . . 9 . . . 8 . . ." He counted down. When he reached 1, Will tried to pull his hands apart and immediately sensed the resistance. In a shock he suddenly dropped the bones to the floor.

"They are magnets!" he exclaimed with eyes wide.

"Of course," I said. "You didn't know bones could be magnetic?"

"But they can't be," Will said, and dashed out of the room.

Brandon and I were chuckling at this, when Will crept back. Before either of us could say anything, Will had grabbed off the floor the two bones he had dropped. He held one of them vertically before his eyes, then touched the second one to the lower tip of the first. He expectantly let go of the bottom one, and, of course, it fell to the floor.

"They aren't magnets," he stated with satisfaction, then ran out the back door.

Brandon and I looked at each another and burst out laughing. Magnetic chicken bones!

Participant Roles

As we move through our description of specific forms, we will introduce several important formulaic qualities of folk illusions' communicative processes, and we begin with a variety of recurring roles necessary for the creation of the intended illusion. These *participant roles* fall into three general categories—participants who facilitate the realization of an intended illusion (*facilitators*), participants who experience the intended illusion (*experiencers*), and participants who observe the performance (*observers*). At the outset, we need to stress the fact that these participant roles—as aspects of social interactions that are always capable of varying in novel ways—represent fluid categories. In the case of Floating Arms, we recognize immediately that the initial performance we described (the variant that involves one child holding another child's arms at her sides) will always require a facilitator (Child 1) and an experiencer (Child 2). Clearly, any performance of this variant may or may not involve observers. For the second and third variants that involve experiencers pressing their arms against the inside of their own pockets or against the inside of a door frame, we see that the experiencers can act as their own facilitator.

Within these broader participant roles, we can also identify specific subvariants across different forms of folk illusions. In the case of the two-person Floating Arms variant, the facilitator instructs and directs the actions of the experiencer, so we refer to the facilitator as the *director* and the experiencer as the *actor*. More precisely, the director provides process-based verbal (and sometimes nonverbal) instructions and physically directs the actor so that the actor experiences a particular illusion—such as the illusion that the actor's arms are floating at her sides. The actor is both the center of any given performance's attention as well as the participant who typically experiences the illusion. When the experiencer serves as her or his own facilitator (as is the case in the second and third variants), the experiencer's actions remain central to the creation of the illusory experience, so we will continue to refer to the experiencer in the second and third variants as the actor.

Morphological Consistency and Ludic Qualities

We utilize the notion of a director and an actor in order to construct a morphology of Floating Arms—a useful, concise representation of the discrete combinatorial units of performance necessary for the success of the illusion:

1a. Initiator(s) introduce(s) performance space/time
1b. Director provides process-based verbal/kinesthetic instructions

1c. Director and actor assume performance positions
1d. Priming period
1e. Director releases actor's arms
1f. Group reacts to the illusion

This morphology includes two heretofore unmentioned units: 1a, the introduction of the performance space/time (that is, the keying of performance), and 1f, the group reaction to the illusion. At this point, we know very little about 1a. Unlike with most traditional children's play, predicting when and where folk illusions will be performed is no easy task. While children are apt to jump rope or play tag on the playground, folk illusions can be performed on the playground or during free time in the classroom, at the park after school, in the lunch room, or during sleepovers in the middle of the night. Likewise, children are not apt to perform a folk illusion on any given day. Unfortunately, folklorists cannot easily predict when or where a folk illusion will be performed.

Because every initiation of a performance will vary, we simply denote a prototypical "initiator" or "initiators"; in reality, initiation of performance may come from any of the participant roles.[6] Regardless, 1a is a necessary performative unit because children do seem to recognize—consciously or unconsciously—while playing with folk illusions that they are inside a time and space where magical, unanticipated feelings and experiences can occur.[7] Our observations have largely consisted of prompting groups of children and young adolescents to play particular types of folk illusions, so while we assume that folk illusions can and do occur on any occasion where children (and sometimes adults) gather for the purpose of "passing the time," the majority of our observations have involved induced natural contexts.[8] In both prompted and impromptu performances, however, it is clear that participants want the illusions to work, and they derive pleasure from the illusion's success. The facilitator receives credit from the group—especially from the experiencer—for introducing a fun activity, and the experiencer becomes an insider to communal understanding of the illusion. While it is probable that children occasionally only pretend they are experiencing an illusion for the benefit of the group, group reactions to the illusion support our claim that experiencing the intended illusion is the central aspect of all folk illusions. In one seventh-grade classroom, students were socially aware of the "best" director for performing a particular type or variant. They might say, "Oh, let Susan hold your arms for you. She's good at it." As we shall see, folk illusions operate within established social hierarchies.

Twisted Hands

Our second example of a folk illusion is an established folkloristic example of a finger game that we call Twisted Hands. During Twisted Hands, the director instructs the actor to twist her hands into an awkward position that makes it difficult for her to recognize which of her fingers is which. In Roger Welsch's gathering of childlore, "Nebraska Finger Games" (1966), the following unnamed activity (our Twisted Hands) is cataloged under the heading of "Visual and Tactual Sensation Games": "The hands are twisted at the wrist, palms down. The palms are brought together and the fingers are folded together. The hands are brought down beneath the wrists and up before the chest. The fingers now lie directly beneath the eyes with the thumbs pointing away from the body. Another player points to various fingers of the folded hands, but does not touch them. The player is challenged to wiggle the finger pointed to. It is surprisingly difficult" (1966, 176). Welsch's explanation is appropriate (see fig. 2.2). Our fieldwork in South Louisiana has shown that Twisted Hands is a living piece of lore with at least one variant (see E4a in the catalog). Along with observing performances of Twisted Hands in southern Indiana, we gathered remembrances of Twisted Hands performances in Illinois, New York, and New Jersey, as well as the Guangdong province in southern China and the Guateng province of South Africa. We find Welsch's morsel of commentary on this example to be simultaneously vague and correct. First, it is indeed surprisingly difficult to move the correct finger in this game; generally, very few move the correct finger on their first attempts. But from what does the difficulty arise?

Experimental study of what we are calling Twisted Hands goes back to the late 1890s.[9] From the outset, observations and experiments have shown that persons usually move the incorrect finger during Twisted Hands. In 1935, Charles Van Riper concluded that the illusion of Twisted Hands exists because of a conflict between visual and tactual inputs. More precisely, what looks and feels like one's right index finger while playing Twisted Hands is not actually one's right index finger. Certainly, Van Riper is correct, and several folk illusions arise from conflict among a given type's necessary visual, verbal, tactile, proprioceptive, and/or somatic inputs.[10] More recently, psychologists and cognitive scientists have attributed Twisted Hands' illusion to multiple (incongruent) representations of an individual's body image and body schema—where body image consists of beliefs about the body's appearance and where body schema is a perception of the body's position,

Fig. 2.2. The actor in this performance of Twisted Hands is pictured away from the director for the sake of demonstration.

movement, and capabilities.[11] Similar to work on the Kohnstamm effect, however, these studies have not considered the folk creation of the illusion.

We have observed that, as a folkloric performance, Twisted Hands involves a director providing verbal and nonverbal process-based directions to an actor. In some instances, the director might be the only participant who is familiar with Twisted Hands, and it might be true that the Twisted Hands' illusion is more potent if an actor is not familiar with the particular type.[12] In Twisted Hands, the director usually demonstrates the correct positioning of the hands and fingers and then points to the actor's finger that is to be moved only after the actor has assumed the correct position. As Welsch correctly notes, the director does not touch the actor's finger while selecting which finger the actor must move, since touching the finger to be moved significantly decreases the potency of the illusion.

For most folk illusions, we choose the term *play* in lieu of the term *game*. We do so, in short, because of the relative sense of inclusiveness between the two terms; it is not necessarily true that all types of play are games, but it is certainly true that all games are—in at least some respects—play. Twisted Hands is a game with governing rules that are necessary for both the continuity of the game and the creation of the illusion.[13] When the actor moves the incorrect finger, laughter, amusement, and some amazement follows. Fun is always a part of folk illusions—which makes them no different, of course, from any other type of play. Children have the most fun playing Twisted Hands when the actor moves the incorrect finger—that is, when the illusion is successful and when the actor loses the game. In Twisted Hands, children are playing with their abilities to manipulate individual embodied experiences through social means. In particular, this form plays with the notion that one can move one's own selected finger if one wants to. Children seem to enjoy the realization that their bodies, in this particular sense, are not givens.

Considering Twisted Hands as a form within the genre, we notice certain parallels with Floating Arms. Particularly, we notice the participant roles of the director and the actor, process-based verbal/nonverbal instructions from the director, performance positions, and the experience and reaction to the experience of an embodied illusion. Twisted Hands' morphology breaks down as follows:

2a. Initiator(s) introduce(s) play space/time
2b. Director provides process-based verbal and nonverbal instructions
2c. Director and actor assume performance positions
2d. Director points at the actor's finger to be moved
2e. Actor attempts to move appointed finger
2f. Group reacts to the illusion

Again, we have yet to observe 2a as what we might call a spontaneous natural occurrence. Step 2a could consist of the director suggesting "Let's play a game" or asking "Wanna see a trick?" But 2a could also be synthesized with 2b if the director immediately asks the actor to put her hands "like this." In 2b, 2c, and 2d, we find the simultaneous transmission of the form and creation of the illusion. It is important to reiterate that the illusion is most potent when the director chooses a finger for the actor to attempt to move. The actor does not typically assume the position of Twisted Hands, look down, and choose his own finger to be moved. The director's pointing is a necessary visual input for the form. Steps 2e and 2f do not occur without the preceding units of performance, and this is another example of a coalescence between aspects of traditionalized performance and necessary inputs for the creation of an illusion. Finally, 2b–2d for Twisted Hands parallel 1b–1d in Floating Arms in that they are both priming periods. The priming period for Twisted Hands is very short, however, because the effect relies on the reversed cognitive (somatic) map of the hands presented in the visual mode. As the actor makes an effort to suppress the visual input and relies more on the kinesthetic map, movement of the chosen finger becomes more accurate. We have observed that the director, and sometimes others in the group, will talk to or yell at the actor, with the effect—intended or not—of inhibiting the concentration necessary to resolve the correct somatic map. Thus, the length and activities of the priming period respond directly to the kinesthetic and cognitive conditions necessary for a successful illusion.

The Chills

With our third folk illusion, the Chills, we introduce the first example of a verbal component that is not process-based. Unlike Floating Arms and Twisted Hands, the Chills' verbal component is lyrical. The Chills also involves heretofore-undiscussed participant roles. For reasons we explain below, we refer to the facilitator in the Chills as the *agent*, and we refer to the experiencer as the *patient*. During a performance of the Chills, the agent stands behind the patient and recites the following lines while performing correlative actions on the patient (see fig. 2.3). We observed this performance while fieldworking with a group of four eleven- to twelve-year-old girls in the summer of 2009:

> Concentration, concentration.
> Crack an egg on your head. [Agent places a fist on the patient's head and slaps the top of that fist with the other hand.]
> Let the yolk run down. [Agent runs fingers down patient's head, neck, and back.]
> Let the yolk run down. [Agent runs fingers down patient's head, neck, and back.]
> Let the yolk run down. [Agent runs fingers down patient's head, neck, and back.]
> Stab a knife in your back. [Agent pokes patient between the shoulder blades.]
> Let the blood run down. [Agent runs fingers down patient's back.]
> Let the blood run down. [Agent runs fingers down patient's back.]
> Let the blood run down. [Agent runs fingers down patient's back.]
> Spiders going up your back. [Agent crawls both hands from patient's lower to upper back.]

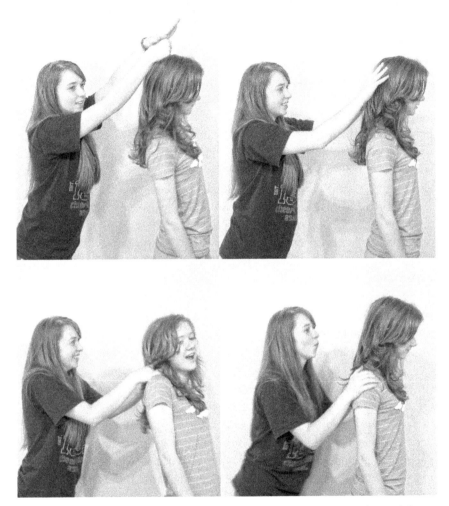

Fig. 2.3. *Top left,* Crack an egg on your head; *top right,* Let the yolk run down; *bottom left,* Tight squeeze; *bottom right,* Cool breeze.

Spiders going down. [Agent crawls both hands from patient's upper to lower back.]

Spiders going up your back. [Agent crawls both hands from patient's lower to upper back.]

Spiders going down. [Agent crawls both hands from patient's upper to lower back.]

Tight squeeze. [Agent pinches the trapezoid area of patient's shoulders.]

Cool breeze. [Agent blows on the back of patient's neck.]

The first time we observed a performance of the Chills, one of the girls looking on exclaimed at the performance's conclusion: "Now you get the chills!"

The participant roles in the Chills differ from Floating Arms and Twisted Hands in that the facilitator's actions primarily create the intended illusions. The experiencer, other than assuming the performance position, plays a very passive kinesthetic and verbal role. We take the terms *agent* and *patient* from their usage in grammatical theories of thematic roles, where the agent in the subject position deliberately acts on the patient in the object position.[14] Take the following grammatical examples: "Tommy chewed his cheeseburger" and "Lydia hugged Ellen." In these two examples, *Tommy* and *Lydia* are agents in the subject position while the *cheeseburger* and *Ellen* are patients in the object position. To be very clear, while the social aspects of touching obviously play a big part in the fun and allure of the Chills, we *do not* mean to imply that this form—or other forms that involve agent/patient participant roles—necessarily involves the contexts of patient and doctor or "playing" patient and doctor.

We have corroborated in our observations that the Chills is a living folk illusion with many variants in Louisiana and Indiana, and we have gathered remembrances from thirteen US states as well as a remembrance from the city of Dubai in the United Arab Emirates and one from Singapore.[15] Sometimes children stab "forks" instead of "knives" into their playmates' backs. Other times the agent chants the refrain "Concentration! Concentration!" between each action. Sometimes only an egg is cracked over the patient's head with no accompanying actions, and other times the agent begins a performance by drawing an X and a row of commas on the patient's back. In the aforementioned Singapore remembrance, the chilling sensations are described as electrical currents and doctors' surgical instruments. Though they do not focus on the perceptual components of the tradition, folklorists Mary and Herbert Knapp briefly describe what we are calling the Chills as a "Scary" in which children force themselves to "cope with the unknown" (1976, 250). We agree. Every performance of this type we have observed serves the purpose of giving the patient "the chills" by coalescing chilling, unknown situations with appropriate, accompanying haptic inputs that result from the agent's physical interaction with the patient.

Let us consider more closely the intended effect of the variant we just described. The first experience of chills results from the patient's feeling the illusion of having an egg cracked on her head and having yolk run down her head and shoulders. Generally, it is true that lightly running one's fingers down another person's neck and back is likely to give that person the chills with or without an accompanying verbal component. For more than a century, psychologists, dermatologists, and neurologists have studied the haptic and emotional nature of knismesis (light tickling).[16] Recent studies suggest that laughter-associated tickling is a social behavior rather than an instinct, and human interactions that involve tickling are highly dependent on social roles of dominance and submission.[17] The agent/patient roles and performance position of the Chills may support those claims. Consider the Chills' morphology:

    3a. Initiator(s) introduce(s) performance space/time
    3b. Agent and patient assume performance positions
    3c. Agent performs lyrical/kinesthetic actions on the patient
    3cc. Patient perceives agent's actions
    3d. Patient gets the chills
    3e. Group reacts to the illusion

Unlike Floating Arms and Twisted Hands, no instructions are provided outside of 3b. Moreover, the performance position of the Chills places the patient in a highly vulnerable position—with her back turned to the agent. Just as all humans react to someone sneaking up behind us, we get something like "the chills" when tickled or lightly touched from behind. In this case, the agent is clearly playing a dominant social role.

Eggs in the Blender

Claiborne Rice and K. Brandon Barker

We closed chapter 1 with a reference to Gilles Fauconnier and Mark Turner's argument that those who seek to understand the inner workings of human minds must first consider mental abilities involved in seemingly trivial tasks like grammar and analogy. Their work, and that of many others since 1995, has examined the ubiquitous human capacity for conceptual blending—that is, the ability to combine two or more ideas together to achieve a new idea. We borrow a couple of Fauconnier and Turner's oft-used examples to explain the basics of their blending theory here:

> Consider a contemporary philosopher who says, while leading a seminar, I claim that reason is a self-developing capacity. Kant disagrees

with me on this point. He says it's innate, but I answer that that's begging the question, to which he counters, in *Critique of Pure Reason*, that only innate ideas have power. But I say to that, what about neuronal group selection? And he gives no answer.

In one input mental space, we have the modern philosopher, making claims. In a separate but related input mental space, we have Kant, thinking and writing. In neither input space is there a debate. These two input spaces share frame structure: There is a thinker, who has claims and musings, a mode of expression, a particular language, and so on. This shared frame structure constitutes a third space, a generic space, connected to both input spaces. There is a fourth space, the blend, which has both the modern philosopher (from the first input space) and Kant (from the second input space). The blend additionally recruits the frame of debate, framing Kant and the modern philosopher as engaged in simultaneous debate, mutually aware, using a single language to treat a recognized topic. The debate frame comes up easily in the blend, through pattern completion, since so much of its structure is already in place in the two inputs. Once the blend is established, we can operate cognitively within that space, which allows us to manipulate the various events as an integrated unit. ([2002] 2008, 59–60)

We have been impressed from the outset of our work with folk illusions by children's abilities to blend semantic inputs and embodied perception into illusory experience, and we find particular resonance with Fauconnier's example of blending that he calls "the Skiing Waiter":

[A] ski instructor was trying to teach a novice (who happened to be French) to hold his arms correctly and look down the slope (rather than in the direction of the skis). The instructor told the novice to imagine that he was a waiter in Paris carrying a tray with champagne and croissants. By focusing his attention on the tray and trying to avoid spilling the champagne, the novice was able to produce something approaching the right integrated motion on the slope. The inputs in this case are the ski situation and the restaurant situation, with arm and body positions mapped onto each other. The generic space has only human posture and motion, with no particular context. But in the blend, the skier is also actually carrying the champagne tray. The blend is of course a fantasy, but it does map the right motion back onto the ski input. Rather remarkably, this pedagogical feat requires no actual champagne and croissants. Just thinking of them is enough. (Fauconnier 2001, 261)

We find it fruitful to think of the illusions created by a performance of the Chills as a kind of conceptual blend, where the patient blends input spaces created by the director's verbal and kinesthetic actions. The first input space comes from the lyrical egg/egg yolk situation correlative to the director's verbal performance, and the second input space comes from the perceptualization of the tickling situation correlative to the director's kinesthetic actions.[18] The generic space includes typical reactions to striking phenomenological experiences like having an egg cracked over our heads, or having spiders crawl all over us, or being stabbed in the back with a knife. The blend produces the reactionary experience of shivery chills.

In their book-length discussion of conceptual integration, Fauconnier and Turner make the point that even though it is possible to identify rather strong constraints, "blending is not deterministic" ([2002] 2008, 53–54). As such, it remains very difficult to predict what blend will arise from any given set of input spaces. In the case of the Chills, however, children seem to have identified a particular circumstance of inputs from which the resulting blend remains quite consistent. Despite the fact that the phrase "the chills" does not always appear in the lyrical variants of the tradition, the children and youths we have talked to commonly report experiencing the chills.

The Chills does not rely on the precarious position of the patient alone. The illusion crystallizes children's instinct that the mind can blend the haptic sensations arising from the agent's actions with the anticipated effect of the verbal component's imaginary situation. Knismesis can and does feel like spiders, but it can also feel like a thick liquid on the skin, like egg yolk or blood. The patient blends the haptic inputs of the agent's actions with the semantic inputs of the agent's lyrics while experiencing the illusion. A performance of the Chills asks the patient to imagine that she actually has an egg sitting on the top of her head. She imagines that when the agent's hand smacks the imaginary egg on top of her head, an egg is actually breaking there. Finally, she integrates her knowledge (or experience) of an uncooked egg's contents so that she can imagine egg yolk is actually running down her head, neck, and back. She can and does imagine these things with ease, and this is an important point. Imagining the egg yolk prepares her to imagine the horrific situation of being stabbed in the back. Assuming that almost all children who play the Chills have never actually been stabbed in the back, we recognize this as an example of highly imaginative play with the body and with illusions. It is a socially derived imagination. The children create the embodied, imaginary situation of being stabbed in the back through a blend of actual physiological and imaginative processes. And as we have already stated, this blending of the haptic inputs with traditionalized, lyrical verbal inputs for the sake of the illusion is indicative of a movement on folk illusions' verbal continuum away from the unadorned, process-based verbal inputs of forms like Twisted Hands toward complex narrative verbal inputs of forms like our last example, Light as a Feather.

Light as a Feather, Stiff as a Board

Light as a Feather is the name we have given a particular variant of a form of folk illusions we refer to as levitation play. Light as a Feather is well known among American youths (especially preadolescents and adolescents). We gathered remembrances of Light as a Feather from eleven US states.[19] Light as a Feather involves somatic components that are fully embedded within an elaborate narrative and imaginary scenario. The full, folkloric name for the variant, "Light as a Feather, Stiff as a Board," refers to folk illusions in which a number (approximately four to six) of participants lift, or levitate, one playmate who is lying on the floor. In its contemporary variants, it is usually characterized by a ritualesque verbal component performed by a participant director that calls for the solemn concentration of all parties involved. Directors often include a narrative verbal component that explains how the actor came to be lying dead on the floor, before prompting other participants to repeatedly chant the words "Light as a feather, stiff as a board, light as a feather, stiff as a board." The illusion of levitation occurs when the participating patient is lifted more easily than would be expected. Most children do not report consciously realizing that Light as a Feather's performance position distributes their playmate's weight evenly among the lifting participants.

Folklorists have studied levitation play among children for several decades now but have rarely examined the participants' embodied experiences. No other folkloristic work seems to have studied the illusory qualities of the form. As to the social characterization of the variant, we must first note that Light as a Feather is typically played at night in a slumber-party atmosphere. This is true for several reasons. First, Light as a Feather plays with the fear of spiritual magic, a fear most palpable at night. Second, slumber parties usually involve the proper number of participants. Third,

Light as a Feather is commonly associated with occult behavior; thus, it must be performed away from any adults' disapproving gaze.[20] Because of these dubious characteristics, folklorists have worked with this variant through vague remembrances, and from those remembrances, they have plucked out the verbal components for the sake of psychological analyses while missing the folk illusion at the heart of the variant.

For example, Iona and Peter Opie include the short section "Levitation" in their oft-cited *Lore and Language of Schoolchildren* (1959) under the category "Some Curiosities." They catalog levitation alongside curious legends, children's remedies for warts, children's beliefs about saying the same thing at the same time (contemporary jinx lore), and so on. The short section on levitation catalogs two witnesses' accounts. We will discuss the Opies' second account first, for it is the oldest. From Samuel Pepys's well-known diary, the Opies cite Pepys's anecdote featuring four French girls and one boy playing with levitation in Bourdeaux, France, in 1665. The anecdote is actually a second-hand account; Pepys recounts the anecdote as it was told to him by one Mr. Brisband:

> He [Mr. Brisband] saw four little girles, very young ones, all kneeling, each of them, upon one knee; and one begun the first line, whispering in the eare of the next, and the second to the third, and the third to the fourth, and she to the first. Then the first begun the second line, and so round quite through, and, putting each one finger only to a boy that lay flat upon his back on the ground, as if he was dead; at the end of the words, they did with their four fingers raise the boy as high as they could reach, and [Mr. Brisband] being there, and wondering at it, as also being afeard to see it, for they would have had him to have bore a part in saying the words, in the roome of one of the little girles that was so young that they could hardly make her learn to repeat the words, did, for feare there might be some sleight used in it by the boy, or that the boy might be light, call the cook of the house, a very lusty fellow, as Sir G. Carteret's cook, who is very big, and they did raise him in just the same manner. (309)

In *Lore and Language*, the anecdote is preceded by the lines the girls are said to have repeated before lifting the boy:

> Voyci un Corps Mort,
> Royde come un Baston,
> Froid comme Marbre,
> Leger come un espirit,
> Levons te au nom de Jesus Christ (309)

Considering the whispering in the ritual, we cannot say exactly how Mr. Brisband recorded these lyrics. Pepys's reference obviously does not include the familiar folkloristic accounts of time, place, and participants. Further, it is not a remembrance, technically speaking, because it is word of mouth. We find no experientially verified information in Mr. Brisband's account. We are less concerned with the actuality of four girls lifting the "lusty" cook than with the fact that the reference does establish the existence of the variant as a part of children's play some 350 years ago. We also find in Mr. Brisband's account several salient features of levitation play still pertinent today. Those are the traditionalized performance position, ritualesque symbolism, several directors lifting an actor, and reference to the actor as a dead, stiff body.

The Opies' other account of levitation play comes from a correspondent who attended school in Bath in the 1940s. The correspondent's account was written on June 22, 1952:

> The queerest happening I know of I have on my own memory. Some of my type-mates one day resolved to try a trick they had heard of, in which one person sat in a chair, and several others (I think four) stood round, with some part of their hands pressed on the victim's head—I think it was their thumbs. The idea was to continue this in an atmosphere of intense concentration by all (no talking or giggles) for a few minutes, after which it was supposed that the person sitting, contrary to all the laws of nature, would actually become lighter and could be lifted by the others as easily as if she were a cushion. This experiment was watched with keen interest by the type, and seemed to have a curiously uncanny effect. Whether by self-hypnotism or not I do not know, but the lifters, with a few fingers only under arm-pits and knees, certainly lifted the seated one with ease and grace into the air, and put her down in the same manner. It was more like real magic than anything else I have ever seen. (Opie and Opie 1959, 309)

This anecdote also shows the salient features of ritualesque order and standard participant roles, but the correspondent's account does not include a reference to the actor as a dead body. In fact, the correspondent's account does not include any macabre verbal component. We find instead another salient feature of the type and the category, the element of concentration. We recall reference to the necessity of "concentration" in the preceding discussion of the Chills. In the call for concentration, children understand (consciously or unconsciously) that they must submit their consciousness to the imaginary situation of the verbal component in order for the illusion to be successful. Though the correspondent's account does not include a ritualesque verbal component, we can still assume that the children were playing within the imagined scenario, "We can levitate our playmate if we concentrate together."

The Opies conclude the "Levitation" section of *Lore and Language* by corroborating the existence of the type with their own remembrances:

> We may add that when we were young we, too, practised this trick, three ways being known to us. In the first, the subject lay down and, if we recollect rightly, there were four raisers using one finger each; in the second, the subject was seated and there were two raisers using a finger of each hand; in the third, which was the most spectacular, the subject stood on a book with his heels and toes jutting over so that fingers could be placed underneath, and he went up vertically. In this last instance the child had to be smaller than the operators or they could not reach the top of his head to press upon. With us it was considered necessary for everyone taking part to hold his breath throughout the operation. One person counted to ten while all pressed downwards, then everyone swiftly changed to the lifting position. (1959, 310)

These remembrances, though generally sparse and vague, corroborate the salient features of the type. In so doing, they begin to reveal possible variants (i.e., different positions for the participants and holding the breath). Ultimately, the Opies' inclusion of levitation play in *Lore and Language* is an important early recognition of the form.

That recognition, unfortunately, has not led to a great deal of fieldwork or research on levitation play. When folklorists since the Opies have discussed levitation play at all, their discussions usually amount to little more than a recognition of a variant alongside other ritualesque, supernaturally oriented folklore that occurs at adolescent girls' slumber parties (see Knapp and Knapp 1976, 252; Bronner 1988, 167). The only scholar who seems to have seriously considered the form is Elizabeth Tucker. In her "Levitation and Trance Sessions at Preadolescent Girls' Slumber Parties" (1984), Tucker analyzes two kinds of supernatural games played by adolescent girls at slumber parties: levitation and trance sessions. Her analysis of levitation play is particularly pertinent.[21]

Tucker's study focuses on data gathered while doing fieldwork with several groups of girls in southern Indiana and New York State between 1976 and 1982. Most importantly, Tucker's article recognizes that the central conscious objective during levitation play is to lift the actor. She explains, "Most of the girls I interviewed or observed doing levitation focused on the act of lifting, with secondary attention to the ritual. If the attempt worked, they were happy, and that was the main thing" (1984, 130). After initially recognizing the embodied experience of levitation play, Tucker's central argument becomes a psychological, symbolic analysis of the variant. Specifically, she theorizes that the salient themes of death and resurrection within the verbal (which she refers to as ritual) performance of Light as a Feather reflect the fact that teenage girls are "keenly aware of the need to reach out: to grow, change, and develop new social roles for themselves" (130). She adds that the ritual aspects of levitation play (especially in the case of the actor) serve as a lesson about their budding adolescent social structures: "Being lifted is not just an experiment with victimization, but also a symbolic submission to peer pressure" (130). This psychological, symbolic analysis is insightful and very well may be correct. If so, Tucker's reading provides us with another example of participant roles being directly affected by the dynamics of social hierarchy.

Tucker also asserts that girls play with levitation in order to "enter another realm of sensation." In levitation play, that realm of sensation, Tucker contends, is the supernatural. She then delineates the chain of reasoning that tells children levitation is a supernatural event: "Lifting a heavy object is hard to do and always takes the whole hand. Two fingers are not enough to lift something normally, so levitation must be extra-normal or supernatural. Empirical evidence, as in so many other areas of children's experience, is of the greatest importance here" (126–27). We find much to agree with in Tucker's description. For more than a century, scientists have recognized the fact that an object's perceived size and volume have a potent influence on its perceived heaviness, and the idea of an inherent mismatch between people's expectations about an object's weight and people's experience of actual, felt weight continues to run through contemporary explanations of weight illusions. Randall Flanagan, who heads the Cognition and Action Laboratory at Queens University, explains that during a size-weight illusion, "the actual sensory information obtained during the lift differs from the expected sensory information estimated from the size of the object" (Flanagan et al. 2001, 90–91). We will look much closer at the issues surrounding weight perception and weight illusions—including the important differences between sensorimotor information and cognitive expectations—in chapter 5, but here we want to highlight the direct connection Tucker establishes between supernatural and phenomenally askew *sensations*. In this connection, the subjective, past experience of individual participants (i.e., sensory experience) merges with communal notions of past experience (i.e., the folkloric tradition).

As was the case with Twisted Hands, Light as a Feather highlights the illusion-inducing dynamics of perceptual and cognitive mismatches. More important for our discussion, however, are the traditionalized components of folk illusions that consistently give rise to such sensory mismatches. Consider, for a final example, Light as a Feather's complex morphology:

4a. Initiator(s) introduce(s) performance space/time

4b. Participants assume performance position

4c. Directing lifter performs ritualesque/magical verbal components (priming period)

4d. Directing and acting lifters levitate patient

4dd. Patient is levitated

4e. Lifters lower patient

4ee. Patient is lowered

4f. Group reacts to the illusion

Compared to Floating Arms, Twisted Hands, and the Chills, Light as a Feather's performance position involves a large group of participants separated into two participant roles: the patient lying on the floor, and a new role, *lifters*, who are divided into a *directing lifter* who provides instructions to the *acting lifters*. The directing lifter, usually seated at the head of the patient, typically leads the verbal performance; she also typically narrates any accompanying stories. Her dominant social role is both actual and symbolic. The patient is in a highly vulnerable position: even more so than in the Chills, because here she is prostrate on the floor, surrounded. We also recognize in 4d, 4dd, 4e, and 4ee that the kinesthetic actions disperse the experience of the illusion among the group.

During a successful performance of Light as a Feather, the lifters anticipate that levitating their playmate with two fingers will be difficult—if not impossible. Because experiences with distributed weight are uncommon—compared to experiences of the self as a sole lifting agent—the lifters have slight everyday experiential basis for categorizing Light as a Feather's actual sensory experience. The same can be said for the patient who lies on the floor, who does not expect that the attempts to levitate her stiff body will be successful. When the patient senses that the lifters are levitating her with ease, she shares with the lifters in experiencing the illusion, but from a different perspective. The additional perspective increases the sense of empirical justification of an actual lift and because the verbal components of the variant have already set the stage for a visit from the supernatural, all of the participants feel as though—they believe—the patient is floating. In this sense, Light as a Feather is a completely intersubjective folk illusion. Youths who have played Light as a Feather often report having been a part of several successful levitations. Light as a Feather's size-weight illusion can and does persist across performances.

Folk Illusions: The Genre

Much has been made of the usefulness of genre theory in the social sciences and in the arts. Folklorists and cultural anthropologists particularly have worried that the presupposition of the very existence of genres limits our understanding of culture. What do we do with folklore that is not a part of any genre (Dundes 1971)? How do we compare ethnic taxonomies with analytic abstractions (Ben-Amos 1969)? How do we separate the artifice of a genre from its natural contexts (Stewart 1991)? How do genres out in the world affect genres within intellectual disciplines (Geertz 1980)? Arguing that folk illusions compose a genre of folklore, we argue for the usefulness of genre theory *and* the actuality of genres in the world.

The four forms of folk illusions in this chapter—Floating Arms, Twisted Hands, the Chills, and Light as a Feather—do gather together analytically. Their shared morphological components of participant roles, performance positions, priming periods, ludic qualities, and perceptual illusions allow us to think at the level of a genre in the context of illusory perception. This is no small feat, for illusory perception has been notoriously difficult to categorize.[22] It is also true that these four forms gather together in ethnographic reports from our youthful consultants. When we show children, adolescents, and our university students a folk illusion like Floating Arms, many remember other activities that created an illusion or an interesting embodied experience. Even Light as a Feather, which we believe creates a size-weight illusion that may be understood as an illusion or as a spiritually magical experience, is mentioned frequently—by the youths themselves—alongside activities like Floating Arms or the Chills.

Alongside genre awareness that we have found in fieldwork and remembrance gathering, consider highly vulnerable performance positions in forms like the Chills and Light as a Feather. It is well established that metaknowledge of ethnic genres is primed (in the cognitive sense) when performance is keyed.[23] The patient's willingness to assume vulnerable performance positions during a performance of the Chills or Light as a Feather grows out of the patient's genre-specific metaknowledge that the vulnerability is necessary for the creation of the illusion *and* that the performance position will not lead to actual, serious bodily harm—even though the ever-present if small prospect of some real harm must contribute to the thrill of the illusion.

In the genre of folk illusions, we also find ample opportunities for the analysis of situated embodied perception, of contextualized knowledge of perception, of artistic ability to manipulate embodied perception, of the performance of embodied perception, and of the communication of embodied perception. In this list, we locate resonance within perennial issues of folkloristics, of children's folklore, and of illusion studies. Richard Bauman, in his oft-cited study *Verbal Art as Performance*, worries that text-centered, etic approaches to verbal art deemphasize the importance of the "association of performance with particular genres" ([1977] 1984, 25). Folklorists have learned this lesson well. We know that competency is always situated. Folk illusions, by definition, defy simplified textualization, and in the absence of readily available texts, we are forced to include embodied aspects of folkloric performance that are always present—such as performance positions—but are not always discussed, even in performance-centered studies.

Notes

1. Portions of this chapter can be found in our original discussion, "Folk Illusions: An Unrecognized Genre of Folklore," which appeared in the fall 2012 issue of the *Journal of American Folklore* 125 (498): 444–73. Used courtesy of the University of Illinois Press.

2. As we will show, priming periods are periods of preparation that are necessary morphological elements for many folk illusions. We recognize that the term *priming* is also used as a technical term in psychology and the cognitive sciences. In these fields, *priming* designates the effects of implicit memory, where one stimulus facilitates or inhibits the processing of a subsequent stimulus. Our use of the term is not necessarily exclusive of these established usages—that is, the intended illusion of Floating Arms is *folk* precisely because the proprioceptive memory fatigue resulting from what we are calling a priming period coalesces with the semantic categories primed—in the cognitive-scientific use of the term— by the verbal characterization of the experience. Children say, "Now, your arms feel like they're floating!"

3. Unless otherwise stated, remembrances were gathered via a survey of university students at the University of Louisiana at Lafayette and at Indiana University Bloomington between 2008 and 2014. Along with remembrances gathered from students who grew up in Indiana and Louisiana, we also gathered remembrances from students who grew up and performed the folk illusion in New York, California, Maryland, Michigan, Minnesota, Wisconsin, Texas, Florida, Illinois, Georgia, Ohio, West Virginia, and Massachusetts.

4. For summaries of physiological studies on the Kohnstamm effect, see Craske and Craske (1986, 123–27) and De Havas, Gomi, and Haggard (2017).

5. See Duclos et al. (2004, 58–59) for a concise review of experiments with mechanical vibration-induced muscle contractions and their aftereffects.

6. For a general discussion about the initiating or "keying" of the related phenomenon of verbal art, see Bauman ([1977] 1984, 15–24).

7. That play is an experience bounded by the expectations of the unexpected is not a new idea. Seminal play scholar Johan Huizinga notes the essential boundaries of play in his

seminal work, *Homo Ludens*: "All play moves and has its being within a playground marked off beforehand either materially or ideally, deliberately or as a matter of course" ([1944] 1949, 10).

8. For a discussion of "induced natural contexts," see K. Goldstein (1967).

9. Henri describes the position of our Twisted Hands in his *Über die Raumwahrnehmungen des Tastinnes: Ein Beitrag zur experimentellen Psychologie* (1898, 83–88). Henri "considered it among other hand-positions in studying the effect of abnormal position upon the motor impulse." In 1904, Burnett named the illusion the Japanese illusion and noted that the subject is usually unable to move any finger pointed to. For a summary of other psychological studies of the Japanese illusion before 1935, see Van Riper (1935, 252–53).

10. For examples, see Zane's illusion (B14), Crossed Fingers (A1), or Double Crossed Fingers (A1b).

11. A summary of this "growing body of behavioral evidence" is in Spence et al. (2004, 185).

12. Van Riper, for example, observed that repeated performances of Twisted Hands eradicated the illusion for nine of ten subjects (1935, 256), but Van Riper's study did not include follow-up experimentation with subjects for whom the illusion had "disappeared."

13. It is certainly the case that all traditional play forms can be performed in ways that vary in the degree that competition is highlighted, but we follow Brunvand in making the distinction between true games and pastimes, in which true games require "the element of competition, the possibility of winning or losing, and a measure of organization with some kind of controlling rules" (1998, 479). In the 1970s, Sutton-Smith argued for possible nuances of distinction in which some games may require only uncertain outcomes and oppositional forces—as opposed to rule-governed winners and losers (1973, 5). In Sutton-Smith's view, activities like Peekaboo can be considered games (see our discussion of Peekaboo in chap. 7).

14. The role of *patient* has not been used in every formulation of thematic roles in linguistic structure. We are generally following Jackendoff's (1987) description, in which the patient is the "object affected" (394). For an introduction to thematic roles (referred to as θ-roles, or theta roles) as a part of government and binding linguistic models, see V. Cook and Newson (2007, 80–86).

15. We gathered remembrances from California, Minnesota, New Jersey, Pennsylvania, Florida, Kentucky, Missouri, Ohio, Illinois, Indiana, Louisiana, New York, and Maryland.

16. Hall and Allin (1897) coined the terms *knismesis* (light tickling) and *gargalesis* (hard, laughter-inducing tickling). Knismesis is usually explained as an evolutionary development that allows mammals to feel insects and parasites crawling on their skin.

17. For a summary and history of scientific studies on tickling, see Selden (2004).

18. Saying that the director's kinesthetic actions create a correlative mental space in relation to the actor's perception of those actions commits us to the point of view that perception—often thought of as a more physiological process—is at least partially constituted by the cognitive, mental operations that help to shape the experience of perception. We borrow the term *perceptualization* as a way to signify the entirety of the perceptual process from ethnomusicologist Cornelia Fales, who deploys the term similarly in her work on timbre illusions (2002, 63–65).

19. We gathered remembrances from Colorado, New Jersey, Pennsylvania, New York, Kentucky, Wisconsin, Illinois, Indiana, Connecticut, Tennessee, and Louisiana.

20. In our fieldwork, we found this to be especially true in southern Louisiana, where more than one individual we consulted refused to discuss the form because it involved playing with spirits and devils.

21. Tucker correctly notes, "The word 'game' does not quite do justice to the nature of what is happening" (1984, 126). Specifically, Tucker contends that certain aspects of levitation play are more ritualistic and less gamelike. This is a complicated distinction to make, for folklorists largely use *ritual* as a term to mark communal events that are governed by a designated, preexisting order and that have clearly defined symbolic elements correlative to any given ritual's morphology.

22. See, for example, Robinson's chapter on the classification of illusions in his *The Psychology of Visual Illusion* ([1972] 1998).

23. Dundes provides the provocative example of beginning the telling of a Christian myth with the traditional frame, "Once upon a time . . . " (1984, 1). In this instance, the Christian would presumably be offended because of the fictional characteristic keyed by the fairy tale's frame.

3

FOLK ILLUSIONS AND THE SOCIAL
ACTIVATION OF EMBODIMENT

FOLK ILLUSIONS OFFER FRESH CONTEXTS FOR THE CONSIDERATION of old problems. The questions of transmission, a good example of an old problem, permeate the study of cultural traditions. In previous scholarly climates when text-centered studies predominated, much effort was spent chasing down the social processes through which a tradition is passed along from person to person and from place to place. Historic and geographic studies of diffusion and polygenesis supposed that a temporal and spatial trace was identifiable and desirable. The focus on performance, however, has shown that transmission must also be considered in the more internalized processes of competency and participation.

The analysis of transmission in a genre that demands attention toward the body significantly challenges any mechanical idea of transmission. If we are all already-embodied people, how can our bodies come to us by any mechanical process? In this chapter, we consider folklorist John McDowell's theory of transmission-as-activation in the context of two heretofore-undiscussed forms of folk illusions, Ping-Pong and What Did I Say? These forms present excellent opportunities to consider what the decentering of the text means for transmission studies because neither form requires verbal characterization. For both Ping-Pong and What Did I Say?, the sociosymbolic kinesthetic actions alone compel illusory perception.

Sections of this chapter appeared in our 2016 article, "Folk Illusions and the Social Activation of Embodiment: Ping Pong, Olive Juice, and Elephant Shoe(s)," published in *Journal of Folklore Research* 53 (2): 63–85.

Ping-Pong at Mountain Bayou

K. Brandon Barker

As with many kinds of children's folklore, fieldworking folk illusions in natural, informal contexts remains difficult. Our initial observation of the folk illusion Ping-Pong, however, did occur in such a natural context, and this is one way that folkloristic methods lend themselves to the study of embodiment. Fieldwork, for folklorists, often refers to commingling with people in their everyday environments, and the folklorist, trained to identify tradition in this context, is best suited to recognize embodied traditions out in the field.

Mountain Bayou Lake Scout Camp, located in Washington, Louisiana, is the prominent campground for the Evangeline Area Council (EAC), Boy Scouts of America. The EAC—one of seven Boy Scout Councils in Louisiana—serves eight parishes in the Acadiana region of South Louisiana.[1] Though the EAC was not chartered nationally until 1924, the Boy Scouts have a long history of active troops in south central Louisiana dating back to 1913.[2] Today, the EAC has approximately 4,500 active Scouts, making it the third largest council in the state of Louisiana. In June 2012, I visited the EAC's Mountain Bayou summer camp for two days and one night, and fortuitously experienced a folk illusion *in media res*.

On the first day of my visit, I spent time going from activity to activity snapping photographs, capturing video, and asking questions. In regimented schedules, different groups of Scouts partook of assigned activities throughout the day. I observed canoeing, motorboating, riflery, first-aid lessons, swimming lessons, outdoor survival skills, leather-working classes, and archery. I saw something else that one expects to see at these kinds of camps. Boy Scouts—ranging in age from ten to seventeen—were engaged to varying degrees in the activities, which meant that some amount of free play occurred.[3] The presence of free play alongside more organized recreational activities is not, of course, surprising. But as I approached the archery range, I saw a form of free play I had never witnessed or heard of before.

To get to the archery range, I needed to walk up an incline from the riflery range, which is located some three hundred yards to the northeast of archery. Approaching the archery range, I noticed two boys about twenty-five to thirty feet away from the demarcated shooting area who seemed to be hitting a ball back and forth using some sort of handheld paddle. The boys looked like they were playing Ping-Pong without a table or a net. The boys not only were hitting a ball back and forth, but they also had an audience of four to six other boys who were all watching and cheering on the players. As the boys hit the ball back and forth, the audience followed the ball's flight, rotating their heads left and right. As I topped the small hill, I could hear the ball colliding with the players' paddles. While jumping and leaning, they were hitting the ball at speeds rivaling Olympic table tennis matches. This feat seemed even more impressive since the boys were playing without a table or a net. But just as my reaction went from being impressed to being truly astonished, I noticed something else very strange about this game of Ping-Pong. There was no ball. In fact, nothing was being hit. The game was an illusion—a folk illusion.

Upon my realization, I immediately began asking the players how the game was played. During a performance of Ping-Pong, each player holds a Styrofoam cup in his dominant, or paddle, hand. The cup is held so that a part of the palm and the fingers cover its open end. The tip of the middle finger rests on the edge of the cup's rim. At the point of a player's swing when his paddle (the Styrofoam cup) would hit the (imaginary) ball, the player flicks the edge of the cup (from the outside in) with the tip of his middle finger. The sound produced by the flick is very much like the sound of a Ping-Pong ball hitting a paddle.

In the heat and humidity of South Louisiana's summers, dehydration is always a concern for campers, so Scouts are encouraged to drink liquids throughout the day. Water coolers and the aforementioned Styrofoam cups are placed at every activity station, and the dining hall serves juices and sports drinks in the cups during and in between meals. The two boys I witnessed playing Ping-Pong informed me that they first learned the folk illusion from fellow campers who were playing across the dining hall at the previous year's summer camp. The boys were twelve and fifteen years old, and both were members of the same troop in New Iberia, Louisiana. Having asked for remembrances from several senior Scouts since that first observation, we have learned that Ping-Pong and some variants have been performed at Mountain Bayou since at least 2005.[4]

Ping-Pong: What Illusion?

What exactly was the illusion experienced during the Mountain Bayou observation? One answer suggests the illusion was the perception of a Ping-Pong ball or some other type of ball when, in fact, there was no ball, but it is difficult to say in retrospect whether one actually perceives a flying ball while observing this folk illusion. Moreover, the ball (or, the illusory impression of a ball) only plays one role in the performance. The illusion, instead, is best understood as distributed across the entirety of the perform-ance, from the mimed actions of the players, to the actual interactions of the audience, to the sounds produced with the Styrofoam cups, and finally to the imaginary ball. Like Light as a Feather, Ping-Pong is intersubjective. The illusion works best when each of the participants plays along.

Consider an actual game of Ping-Pong or some other Ping-Pong-like game (e.g., paddle ball or tennis). The salient physical components of the game are two players, a ball, a handheld instrument with which to hit the ball like a paddle or a racket, a table or court establishing game boundaries, and possibly an audience. Ping-Pong, the folk illusion, clearly accounts for each of these. It also accounts for the salient kinesthetic actions of the game, or what we call the performance positions: the placement of two appropriately spaced, facing players; the swinging of the paddles and the resulting back and forth of the ball; the players' reactions to the physics involved with the flight of the virtual ball; and the audience's tracking of the ball's flight. Ping-Pong also accounts for the dynamic emotional reactions to the competition. As players win and lose (illusory) points, they display the kinds of positive and negative reactions we would expect from players winning and losing actual points.

The most vivid parallel with an actual game of Ping-Pong remains the iconic sound of a ball colliding with a paddle—the onomatopoeic source for the game's name—that is produced by the flicking of the Styrofoam cups. Like most performances, Ping-Pong plays with humans' reliance on inputs from several different modes of perception (i.e., vision, touch, sound, etc.) in order to make sense of the world. The insistence on multimodal (or multisensory) integration was a cornerstone of early twentieth-century Gestalt psychology, and the past few decades have shown a renewed interest in the experimental study of multimodal perception by cognitive scientists and psychologists.[5] Recent research has continued to chip away at structuralist paradigms that divided the processes governing separate sensory modalities. Psychologist Lawrence Rosenblum summarizes these changing notions of multisensory perception in his sweeping book on the senses, *See What I'm Saying*: "The long-held concept of the conceptual brain being composed of separate sense regions is being overturned. Your brain seems designed around multisensory input, and much of it doesn't care through which sense information comes. Your brain wants to know about the world—not about light or sound, as such" (2010, 280). Taking the world of the Ping-Pong performance in its entirety, we find that the sound created by flicking the Styrofoam cups becomes only one small part of experiencing the performance as a whole.

One particularly relevant recent finding for our analysis of the folk illusion Ping-Pong is the so-called flash-beep illusion discovered by Ladan Shams and her team of psychologists at UCLA:

> We have discovered a visual illusion that is induced by sound: when a single visual flash is accompanied by multiple auditory beeps, the single flash is incorrectly perceived as multiple flashes. These results were obtained by flashing a uniform white disk . . . for a variable number of times (50 milliseconds apart) on a black background. Flashes were accompanied by a variable number of beeps, each spaced 57 milliseconds apart. Observers were asked to judge how many visual flashes were presented on each trial. . . . Surprisingly, observers consistently and incorrectly reported seeing multiple flashes whenever a single flash was accompanied by more than one beep. (Shams, Kamitani, and Shimojo 2000, 788)

Vision has historically been considered the dominant perceptive mode for humans.[6] The flash-beep illusion, though, demonstrates that auditory cues can influence visual perception, even when the visual stimulus is not ambiguous.

Ping-Pong constitutes an example of a folk illusion where auditory perception guides or dominates visual perception. Ping-Pong adds important new elements to consider regarding scientific studies of such illusions as well. The social context and traditionalized elements of Ping-Pong are both essential to a performance of the form and necessary elements of the intended illusion's eliciting conditions. Ping-Pong depends on the participants' and the observers' past experience with games of the kind. Compared with Shams's beeps and flashes, Ping-Pong is a more culturally conditioned illusion.

Ping-Pong: Whose Illusion?

Ping-Pong raises another important question about the genre of folk illusions: Who is meant to experience Ping-Pong's "intended embodied illusion"? If an onlooker who has not seen such a performance in the past believes that he is witnessing an actual ball being hit back and forth, then the answer seems self-evident, but how often does this actually happen? In truth, we cannot know how frequently a naive observer of a performance of Ping-Pong believes she is seeing a ball being hit back and forth, but tricking an onlooker into perceiving an actual ball does seem to be the point of the performance on at least some occasions.

The intended experiencer of an illusion, in a theatrical performance such as Ping-Pong, may or may not be the actual experiencer of the illusion. Our understanding of participant roles needs to be variable. For the performance witnessed by Barker at Mountain Bayou, for example, we could construct a morphology like this:

5a.   Initiator(s) introduce(s) performance space/time
5b.   Directing player one begins game with initial swing and hit
5bb. Directing player two returns player one's serve
5bbb. Directing players continue Ping-Pong match
5c.   Acting observers (i.e., the Ping-Pong match's fans) notice and react to the directing players' performance
5d.   Patient (i.e., Barker) observes match and experiences illusion

Given that Barker was the patient in this example, we can only speculate as to the specifics of 5a–5c. Clearly, there are many possibilities for how a performance of Ping-Pong might never reach 5c or 5d. We have witnessed at Mountain Bayou that certain youths practice flicking the Styrofoam cup in order to improve the mimetic quality of the sound, so a given instance of producing the sound does not necessarily key a full performance of the game 5a–5bb.

Online manifestations and representations of related performances corroborate the variability of participant roles. In November 2012, a search of "Ping-Pong no ball" on YouTube yielded videos of a few performances of Ping-Pong like the one at the camp. In example 1, two early teenage boys perform Ping-Pong across a stainless steel table using Styrofoam plates instead of cups to make the noise (Lamar 2011).[7] In example 2, two boys seemingly in their late teens perform a similar game by flicking the backside of actual wooden paddle-ball paddles (Darth Thanos 2010). Notably, this performance ends when one of the boys feigns hitting the ball too high. The imaginary ball flies over his opponent's head and is lost, thus ending the game. In these examples, we quickly recognize the absence of acting observers (except in the sense that the acting observers are the YouTube viewers) and the absence of any patient observer.

In a third online variant of Ping-Pong, however, we see a performance that tricks a specific individual—where the directors appear to perform for a designated patient.[8] The dimly lit video shows two young men (directing players) performing Ping-Pong while standing on opposite sides of a narrow street. The street runs adjacent to what appears to be a pub with outdoor seating. When a pedestrian (the patient) passes by the performance, the young men pretend they have missed and lost the ball. One of the directors comments, "Oh, it's over there, it's over there." He proceeds to ask the pedestrian if she can help them find the ball. The pedestrian looks around quickly. She replies, "Well, I would if I could see it. It must be a very small ball. I can't see it, sweetheart, sorry." As the patient walks away, the directors and several onlookers at the pub break into laughter. Though it is not clear that the illusion has truly succeeded in tricking the patient—she might have cannily integrated herself into the performance space of the game—the performance still provides a good bit of fun for the group.

How, then, should the success of a performance of Ping-Pong be judged? We have previously proposed that fun is always a part of folk illusions. Does the fun had by the directors and the onlookers provide evidence that the performance of the folk illusion was successful? Should success be considered in varying degrees? These questions directed at an explanation of the folk illusion Ping-Pong point us toward central issues in the study of the genre. Specifically, they force us to reconsider our notion of facilitators and experiencers. If we understand the patient to be the intended experiencer of the illusion, then in the case of Ping-Pong, we find that sometimes a patient is not present. Performances can and do occur with the participation of only two directors. At other times, performances of Ping-Pong occur with the participation of two directors and a variable amount of actors—watching the game's back and forth. These actors, like the observers witnessed at the camp, may or may not perceive an "actual ball" being hit.

Regardless, Ping-Pong satisfies its purpose of being enjoyable to these "fully aware" directors and actors, for even when no unsuspecting patient is present, directors and observers remain interested in the illusory game. The Scouts cheered and laughed as they watched the imaginary ball fly through the air. The performance was real. The competition inherent in that performance was certainly real. The directors manipulated the emergent physics attendant to the illusion in order to make the game more difficult for the opponent. One director, for example, swung his "paddle" with great velocity in an upward fashion only to look up into the sky as though the ball were gaining altitude in a way that would be impossible to achieve in a real game. His opponent, in this instance, was forced to stumble around searching into the skies for the falling ball. Both actors' gazes followed the heavenly trajectory. The game only stopped when one of the overzealous directors flicked his Styrofoam cup with too much force, breaking the cup. In defeat, the Scout fell to his knees as the actors cheered for the victor.

Here, we see that a well-performed director role in a performance of Ping-Pong is sufficiently illusion-inducing that the performance remains fun even for participants who are fully aware that no actual ball exists. The social and perceptual inputs are so powerful that participants willingly deprecate in awareness any indications that might pierce the bubble of their illusion.[9] The knowledge that no ball exists does not stop the participants from acting, or feeling, as though a ball exists. We argued in chapter 2 that the traditionalization of folk illusions already implies successful experiences of the intended illusion, and Ping-Pong furthers this argument by showing how knowledge of the fact that an illusion is not "real" does not stop the illusion from maintaining its central qualities of fun and wonder. Rather, the fun and wonder themselves contribute to the perseverance of the illusion despite apparent detractors. The pleasure derived from a performance of Ping-Pong in the case of fully aware directors and actors results from the folk illusion's ability to allow participants to hover between the worlds of illusion and reality.[10]

Not unlike Antonin Artaud's call for a theatrical return to *true illusions*, Ping-Pong reveals the communal quality of baselessness—which is to say that the unrealities we make together always reflect the reality of presupposition. Inasmuch as folk illusions pierce the veneer of a perceived materiality, the genre resonates with Artaud's famous, theatrical calls for "a reassertion not only of all the aspects of the objective and descriptive external world, but of the internal world, that is of man considered metaphysically" ([1938] 1958, 92). In his ethnographic study of a kind of folk drama, *All Silver and No Brass: An Irish Christmas Mumming* ([1975] 1983), Henry Glassie describes landing on Artaud's work in his search for Western commentary on authenticity: "Like the later works of Joyce, Kandinsky, and Levi-Strauss, Artaud's new theater was conceived in opposition to the materialistic, empirical realism of the nineteenth century. The new theater will not be built of endless dialogues in which psychological mysteries are reduced to facts; it will unfold in music and dance, poetry and humor, in improbable symbolical violence. It will be a theater of delight, of vague, profound terror" (65). Illusions—though they are in one sense materially unreal—cannot be anything other than authentic, for they stem from the very systems of perception that facilitate subjectivity. Illusions, the child learns, are both subjective and social, as is reality itself. If—as Chinua Achebe and Henry Glassie suggest—darkness holds a vague terror, profundity both relieves and reinvents the terrible.

In the most socially complex variants of Ping-Pong that involve two directors, a set of acting observers, and an observing patient who experiences the illusion of an invisible ball, we find that the responsibilities of facilitation are spread across the actions of both the directors who pretend to hit the ball back and forth *and* the acting observers who assist the directors' performance by tracing the flight of the imaginary ball. This communal nature of facilitation in Ping-Pong provides a good demonstration of why participant roles—in relation to the communal creation of an illusion—can only be fully understood in the context of a given performance. From the perspective of the acting observers, the Ping-Pong players (the directors) *are* facilitating the actors' experiences, but from the patient's perspective (when a patient is present) the Ping-Pong players and Ping-Pong fans work together to facilitate the illusion. The emergence of participant roles is fluid.

What Did I Say? I Love You, Elephant Shoe, and Olive Juice

Unlike Ping-Pong, the folk illusion I Love You did appear in the provisional catalog of our introductory discussion of folk illusions (Barker and Rice 2012). Therein, we described two variants of I Love You in the section "Visual Illusions":

> a. Elephant Shoe—The agent, after finding an appropriately intimate moment during which she is making eye contact with the patient, mouths the words "elephant shoe." For the patient, the agent appears to be mouthing "I love you" (personal remembrance).
> b. Olive Juice—The agent, after finding an appropriately intimate moment during which she is making eye contact with the patient, mouths the words "olive juice." For the patient, the agent appears to be mouthing "I love you" (remembrance gathered from T. Mitchell, age 32, summer 2011). (465)[11]

Though we still have not witnessed a naturally occurring, nonprompted performance of I Love You, since 2011 we have gathered several remembrances of I Love You from adults and teenagers alike, so we are convinced that I Love You exists as a widespread folk illusion that is performed by contemporary youth in the United States. Because I Love You is usually performed in somewhat intimate settings, we may never witness a non-prompted performance not of our own making. Regardless, tradition and the folkloristic method of gathering remembrances provide us a way to examine the form.

We know that a performance of I Love You occurs in an instant. And while we suspect that I Love You is prototypically performed in social situations where the agent's performance may be readily (mis)interpreted as a romantic advance, we also learned through remembrances that I Love You is sometimes performed among platonic friends.[12] Whether or not any given performance of I Love You is performed among romantic acquaintances, it is also likely that I Love You is popular among teens and early teens simply because love is in the air. Given the biological facts of puberty and the intensity of teenage social networks, we are not surprised that adolescents play with the most important of romantic phrases.[13]

I Love You's intended illusion hinges on mistaken lip reading on the part of the patient, but the remembrances we collected suggest that the illusion may be most intense when romantic interests between the patient and the agent are possible. The patient will be all the more willing to imagine that he has understood the silent message and thus all the more disappointed as the agent explains with a laugh, "I didn't say I love you. I said Olive Juice." I Love You highlights the fact that folk illusions are always performed within the context of preexisting social dynamics. Growing up, we know, can be cruel.

Just as studying the intended effects of many other folk illusions offers insight into the nature of specific illusions, the lip-reading illusion of I Love You offers a new perspective on lip-reading illusions as they are usually studied and understood in the experimental sciences. Lip reading, in fact, has garnered much attention in illusion studies over the past several decades. The most widely recognized of these phenomena is known as the McGurk effect, named for the lead scientist whose team published its findings on the illusion (McGurk and MacDonald 1976). Working with nearly one hundred subjects, including children and adults, McGurk and his colleagues found that speech receivers perceived illusory sounds when presented mismatching visual and auditory inputs: "It stems from an observation that, on being shown a film of a young woman's talking head, in which repeated utterances of the syllable [ba] had been dubbed on to lip movements of the syllable [ga], normal adults reported hearing [da]. . . . When these subjects listened to the soundtrack from the film without visual input, or when they watched untreated film, they reported the syllables accurately as [ba] or [ga]" (McGurk and MacDonald 1976, 746). Since McGurk introduced this phenomenon, hundreds of subsequent studies have confirmed its existence. Rosenblum summarizes the importance of the McGurk effect by establishing its effects on contemporary notions of multisensory processing: "The McGurk Effect is one of the most referenced phenomena in speech perception research. The effect demonstrates that we all lip-read and use visual speech information. It shows that our use of this information is immediate, automatic, and to a large degree, unconscious. It shows that despite our intuitions, speech isn't just something we hear. More generally, the effect, along with thousands of studies it's helped motivate, has shown that our brain is designed around multisensory input. In an important way, your brain doesn't care through which sense it gets its perceptual information" (2010, 246). A performance of I Love You plays with the very fact that visual speech inputs can combine with appropriate social contexts so that the patient cannot help but perceive the words *I love you* coming from the agent's lips.

Alongside work with the McGurk effect, we find a good deal of experimental data on lip-reading more generally. Even though experimental scientists do not yet agree exactly how humans integrate visual and auditory inputs while processing speech, more and more evidence shows that multisensory processing—including auditory, visual, and sensorimotor processing—is integral to speech reception.[14] This fact rings true in a performance of I Love You, since a very specific linguistic message is delivered without any auditory input. I Love You, like Ping-Pong, does not rely on the descriptive semantics of a verbal component to characterize or to assist the creation of the illusion, but unlike Ping-Pong, we cannot say that a performance of I Love You is not linguistic. Rather, I Love You is a linguistic illusion—aimed specifically at creating the sense of an auditory input, but silently.

Folk Illusions and Shared Body Awareness

An unnamed variant of a performance of Ping-Pong that sometimes occurs at the Scout camp is actually an interruption of performance. In this variant, which we learned about via a remembrance from EAC Eagle Scout B. Dale (age eighteen) in the summer of 2013, an observer of a performance of Ping-Pong breaks into the performance space and steals the imaginary ball—often running away with the ball and possibly passing it along to other spectators. Of course, taking a ball away so that play cannot continue actually occurs when real balls are in play, and stealing the ball is an important objective in several organized forms of play like soccer, basketball, and American football. In American English, a sore loser is traditionally referred to as someone who wants "to take his ball and go home."

Cultural precedents for "taking the ball" guide this variant of Ping-Pong, but *why* spectators sometimes take the illusory Ping-Pong ball is less important for our analysis than *how* the spectators steal the ball. Taking the ball from Ping-Pong's central directors requires that the thief integrate himself into the directors' established illusory performance space. The thief cannot simply throw his hand into the air and exclaim with a laugh, "I've taken your ball!" In order to break the performance successfully, the sneaky spectator must appropriately time his illusory theft with the ball's flight path as presented by the kinesthetic actions of the directors. This is no different, of course, from what would be required of the spectator who wishes to steal an actual ball being hit back and forth.

Taken as a variant of Ping-Pong, traditional performances of stealing the ball at Mountain Bayou present important evidence for how folk illusions compel intersubjective participation. Like the prototypical performance of Ping-Pong, the stolen-ball variant does not work via a coalescence of somatic and verbal inputs. Spectators need not explain to the directors (or to themselves) that they are timing the actions of their theft with the established performance space; instead, they simply act in accordance with the intersubjective context of the illusion. Ping-Pong opens an (illusory) perceptual space that the performers and the spectators feel compelled to join.

The same can be said for a similarly powerful variant of What Did I Say? that we call Fuck You. In a performance of Fuck You, the director menacingly mouths the word *vacuum* toward an intended actor who perceives the director as mouthing the words *fuck you*. Like the phrase *I love you, fuck you* is a specially marked verbal phrase—especially for youth and teens. Where *I love you* signifies all of the complicated aspects of romantic feelings that youths experience, *fuck you* signifies the social discord of adolescence. *Fuck you* is also specially marked for youth because the phrase is almost always identified by adult supervisors as a curse word, and its usage, if discovered by an adult, may result in punishment.

Like Ping-Pong, the "vacuum" variant of What Did I Say? has a small but recognizable online presence. At the time of this book's publication, "vacuum" is defined on urbandictionary.com as follows: "A word that, mouthed, looks like 'fuck you.' Use if you want to mess with someone, or if you're too much of a wimp to say the real thing."[15] The definition aligns with the following remembrance we recently gathered from Meghan Smith (age twenty-four): "I was twelve or thirteen at the time, living in Provo, Utah, and was in a middle school (seventh and eighth grade) context. I was sitting at a table in a crowded, noisy lunchroom with a group of girls. When I saw Stephanie, my middle school enemy, across the room, my neighbor suggested that I mouth the word 'vacuum' to her. At first, I did not understand, and the girl told me to note what it *looked* like she was saying when she did it. I was slow; it took me a while to figure it out." Meghan's remembrance highlights the agonistic qualities of this folk illusion. A middle school enemy is spotted, and youths collectively devise a safe way to attack that enemy "verbally."

Meghan's remembrance also gives us a glimpse at how transmission of such a folk illusion takes place. Not unlike the stolen illusory Ping-Pong ball, Meghan's remembrance actually recounts a somewhat stunted performance. In this case, her enemy, Stephanie, was not the audience of the performance. The enemy only served as a catalyst; then, Meghan struggled to interpret her friend's performance. Her friend, however, trusted the power of the tradition. Without overexplaining, she guides Meghan's reception: "At first, I did not understand, and the girl told me to note what it *looked* like she was saying when she did it." Meghan's narrative offers us a glimpse at two central aspects of the folk knowledge of our bodies as evidenced in performances of folk illusions: (1) The traditionalization of folk illusions can be taken as evidence of their compelling nature. (2) Just as folk illusions are compelling, they represent a specially marked cultural, folk awareness of our body's perceptual systems.

Meghan's friend, who wants to help her with a social circumstance in which an enemy needs to be confronted, does not come right out with an explanation. She does not say, "Look at me mouth 'vacuum,' it looks like I am saying 'fuck you.'" Her directions work instead via example and a strong conviction that the entire folk group's perceptual systems will successfully interpret such a performance. As we hinted above, Fuck You, I Love You, Ping-Pong, and Ping-Pong's interrupted variants all work this way. The respective illusions are created from within the otherwise "given" systems of language and perception, so for youths who experience a successful performance, the salient characteristics of the illusion take their place within those given systems. In other words, performers can rest assured that onlookers cannot resist (to some degree) joining the experience of the intended illusion.

The inherent understanding of embodied processes that coincide with performances of folk illusions like What Did I Say? and Ping-Pong signals an impressive epistemological and developmental feat. As youth learn about different folk illusions, they expand their understanding of their own and of others' bodies. They come to realize—though we certainly would not expect them to articulate the realization this way—that their own and their playmates' perceptual systems can be systematically manipulated to perceive unrealities or unanticipated realities. They learn, intuitively, that offline embodied processes constitute an important part of the whole of experience.

Folk Illusions and the Activation of Bodily Awareness among Children and Adolescents

The notion that knowledge emerges or is created as a part of folkloric communication is closely related to ideas about the transmission of folklore. As we mentioned at the beginning of this chapter, definitions and analyses of transmission often include a transference of some sort of knowledge between subjects (e.g., a folkloric text or the wisdom associated with a folkloric text). This is true for bodily knowledge. In the folkloristic study of material culture, where the focus is on creative individuals and where processes can be observed, the issue of transmission has long been formulated in line with performance theory into a categorical sequence for analysis: learning, development, practice, creation, communication, consumption, and teaching. In these cases, the body becomes a kind of material constant that makes observable, for example, how one potter teaches another how to throw clay.[16] Transmission is rarely studied, however, alongside epistemologies that result from the timely intersection of bodily development in childhood and particular folk traditions of childhood. What can we make of transmission and the epistemic qualities of folk illusions? Our answer comes via a powerful alternative approach to "transmission."

In his "The Transmission of Children's Folklore," John McDowell suggests that the vastly varying and constantly creative world of children's folklore does not fit within the descriptive boundaries of transmission—where transmission posits "images of a superorganic process, a perpetuation of 'items' with their peculiar 'life histories,' quite external to the everyday communicative exchanges of ordinary human beings" (1999, 61). In its place, McDowell suggests a new term, *activation*, for the interchange of folklore: "These considerations lead to the suggestion of a neutral term, perhaps the activation of children's folklore, to refer to the processes set in motion as traditional competencies enter into finite communicative settings among children. Within this constellation, transmission intact or in recognizable variants would remain as one possible outcome, but the folklorist would be alert to the creative, transformative potential of all such encounters" (62). Similarly, Katharine Young, who has worked most extensively with bodylore as a subject of folkloristics, argues in her analysis of gestures, "We activate folklore by moving into it bodily, by lending things our bodies, by borrowing other's subjectivities" (2011, 82). Where Young's usage of the idea of activation metaphorizes intersubjectivity and folkloric communication, McDowell's usage outlines conceptual categories for creative activity within the spaces of tradition.

For McDowell, there exists uncountable opportunities for the child to participate successfully in the folkloric performance without clearly recreating that which is marked as traditional: "Children's folklore performances amply exhibit the effects of both rote memorization and improvisation on the basis of traditional models" (1999, 56). As an example, McDowell highlights the structural and thematic requirements of a riddling session he witnessed while working on Richard Bauman's Texas Children's Folklore Project in 1974. Therein, in one performance, two boys and two girls present riddles that are known to be traditionalized like "What did the rug say to the floor?" alongside emergent, nontraditionalized riddles like "What did the dead penguin say to the live penguin?" The latter riddle, though not traditional, adheres to this "What-did-X-say-to-Y?" category of traditionalized rules for riddles:

1. The question specifies two entities in conversation.
2. Neither of these entities is normally included in the category of speech participants.
3. A motivation for dialogue must exist, either in the form of a shared identity (little chimney, big chimney) or habitual proximity (rug and floor).

These rules exist implicitly within the exemplars of riddles that participants experience during performance; simultaneously, these rules reappear, to varying degrees, when performers activate fresh, analogous conceptual spaces to subtend new exemplars.

Likewise, Ping-Pong and What Did I Say? get transmitted by an explicit performance of the respective forms *and* by an implicit activation of the related conceptual and perceptual spaces. Unlike the activated conceptual spaces of "What-did-X-say-to-Y?," riddles that are governed by the structure and semantics of preceding verbal performances, folk illusions' activated variants are largely governed by the interaction between the supplied cultural contexts (e.g., the actual game of Ping-Pong or the visual presentation of linguistic articulation) and the perceptual systems of the participants. To create new, successful folk illusions, children follow the lead of their playmates into themselves. In this way, the child's body is both discovered and created.

This is to say that we agree with both McDowell and Young, and we add that in the child, folk illusions activate a kind of consciousness of the system of perceptual processes that shapes both the traditional and the creative forms of play. For the observing child, those perceptual processes guarantee his participation in the performance of Ping-Pong, and as the observer becomes a disruptive director, he does so within the preexisting "rules" of the form. Awareness activated by the perception of the illusion allows for variation—even if the variation directly inhibits the initial performance. Likewise, when youths come to see each other mouthing illusory words back and forth, they recognize the possibility for the trick to be employed in varying contexts and for varying purposes—for love, for hate, and possibly for everything in between.

As an epistemological force, folk illusions play with a simple but powerful fact of human perception: The singular embodied subject never fails to categorize perceptual experience. Sometimes categorization leads to veridical perceptions; other times categorization leads to illusory perceptions. Moreover, the complement of veridical and illusory perceptions allows for intuitive comparisons that help make awareness of perceptual processes possible. From an ethnographic point of view, folk illusions—like singularly subjective illusions—are systematic, robust, and surprising, and just as the subjective illusions of the individual give access to the inner workings of embodied perception for the individual, folk illusions provide access to the inner workings of embodied perception for the group.

Notes

1. One could argue that Boy Scout summer camps, which occur annually and are formally marked by traditionalized rituals, customs, and uniforms, do not constitute the everyday. From a folklorist's perspective, that may be true, but we are referring here to the differences between a natural performance of Ping-Pong as compared to a prompted performance out in the field or as compared to a controlled performance in the experimental scientist's lab.

2. The first Boy Scout troop in the areas now served by the EAC was New Iberia's Troop 1, organized by the First Methodist Church of New Iberia in 1913. The second, and more

influential, group was Franklin's Troop 1, organized by the First Methodist Church in Franklin, Louisiana, in 1914. Troop 1 of Opelousas, Louisiana, was also formed in 1914.

3. That is, free play accompanied every activity, excluding riflery, which consisted of highly organized small-group sessions.

4. In the summer of 2013, Barker worked with Eagle Scout Brandon Dale, who informed him that he had performed and witnessed Ping-Pong performances over the years in which he attended the camp starting in the early 2000s.

5. Bertelson and De Gelder's excellent chapter "The Psychology of Multimodal Perception" (2004) reviews and considers multimodal perception historically while attending to important theoretical approaches and categories of research.

6. For example, Aristotle puts vision at the top of his hierarchy of the senses, calling vision "the most highly developed sense" (*De Anima* 429a). As specific examples of sight dominating or overruling other perceptual modes, Shams, Kamitani, and Shimojo cite the well-known examples of the ventriloquism effect, in which the perception of speech sounds seems to come from a source other than the actual speech producer, and of visual capture, in which the direction of the movement of a particular sound is mistakenly aligned with the direction of an associated visual component.

7. While this performance is not nearly as dynamic as the performance of Ping-Pong that Barker witnessed, we are impressed by the boy's similar, formularized use of Styrofoam.

8. This performance was accessed on September 15, 2015, on YouTube, though it has since been removed from the website. Nonetheless, there remain available several videos of illusory Ping-Pong matches performed in order to trick and beguile onlookers. We are fond of variants called "penny pong," during which performers pretend to hit a penny back and forth using waxed paper cups (see, e.g., Dormtainment 2011).

9. The allusion here to Coleridge's "willing suspension of disbelief for the moment, which constitutes poetic faith" is deliberate. See A. Cook (2006) for a summary of treatments of the cognitive poetics of suspension of disbelief in literary art and her proposal for considering it an example of "living in the blend," as explained by Blending Theory. As with the application of Blending Theory to the Chills in chap. 2, we suggest that participants actively blend the abstract organizing schema of Ping-Pong with the ongoing events of the performance. In Fauconnier and Turner's terminology, the ball would be part of the emergent structure of the blend.

10. We thank our reader, Katharine Young, for this astute analysis of Ping-Pong and the pleasure of its experience.

11. Some have suggested that the tricks of What Did I Say? might manipulate the actions of a naive director. In these cases, the director would be instructed to mouth a seemingly innocent phrase like "elephant shoe" at a playmate without knowing that the playmate will perceive the communication as the phrase "I love you." The director in this case would have unwittingly made a romantic pass at the actor. While we grant that this variant is entirely possible, we have not observed or collected a remembrance of such a variant. Taken together, forms of What Did I Say? undoubtedly represent a kind of prank, which we previously labeled Torturous Illusions (Barker and Rice 2012, 468–69).

12. Remembrance gathered from S. Hesse, age twenty-five. She recalled that she and her other female friends often performed this illusion among one another in a platonic fashion.

13. Though not as extensive as work on adult romantic relationships, a large amount of work has focused on adolescent romance and adolescent romantic relationships in the past half century. See *The Development of Romantic Relationships in Adolescence* (Furman,

Brown, and Feiring 1999) for a sweeping introduction. Therein, see especially a discussion of emotion in adolescent romance for an interesting analysis of the ways that romance can "thrust an adolescent into hyperspace: where things are turned upside down, ordinary reality recedes from view, and he or she leaves the envelope of routine daily life" (42).

14. For an easily accessed introduction to lip-reading studies, see Rosenblum's popular book on the subject, *See What I'm Saying: The Extraordinary Powers of Our Five Senses* (2010, 247–56). For a recent scholarly article that provides a summary of pertinent theories (especially neurological theories) along with a robust bibliography, see Okada and Hickok (2009).

15. See Urban Dictionary, s.v. "vacuum," accessed September 15, 2015, http://www .urbandictionary.com/define.php?term=vacuum. For a second entry relating to this variant of What Did I Say? listed as "I want a vacuum," see Urban dictionary, s.v. "I Want a Vacuum," accessed September 15, 2015, http://www.urbandictionary.com/define .php?term=I%20Want%20A%20Vacuum. Of interest also is a version of Cee-Lo Green's song "Fuck You" performed as "Vacuum" by a young woman apparently in a school setting (adanatorememberful 2012). She begins the song timidly as an adult is in the background, possibly within hearing distance, then she gathers strength as he moves away.

16. Ethnographic, situated examples of material culture studies best exemplify the processes of transmission and tradition. For excellent examples, see Jones (1989, 253–63) and Glassie and Shukla (2018, 105–47).

4

FOLK ILLUSIONS AND ACTIVE
PERCEPTION

FOLKLORISTIC WORK WITH CULTURAL TRADITIONS AS LIVED, OUT-IN-THE-WORLD experience serves the opposite purpose of the scientist's lab. It complicates. Dorothy Noyes expressed the paradox of problematizing data sets when she described folklorists' focus as a social base that remains cumbersomely situated between "core binary oppositions of Western modernity: old and new, particular and universal, fluid and fixed" (2012, 14). Paradox presents its own kind of illusions by forcing seemingly impossible thoughts into consciousness, and even though Noyes's list might be more easily understood as old *or* new, particular *or* universal, fluid *or* fixed, the less-illusory "or" adds no clarity because *paradox* more clearly represents the reality of folkloristic work. Our argument that folk illusions spread from group to group, from person to person, by activating communal awareness of embodied sameness reifies folkloristics' enigmatic focus because it raises the specter of *differentiation*—another old, paradoxical problem.

The problem is this: If the transmission of folk illusions works by way of embodied similarity—the shared tendencies for experiencing certain kinds of illusions like the Kohnstamm effect, size-weight illusions, or illusory knismesis and gargalesis—how and when do the processes of perception allow for variation of folk illusions' aesthetic and communicative qualities? How do the hallmarks of folkloric tradition like volition and localization meld with perceptual illusions? What constitutes a folk illusion's tradition in the first place—or as Elliot Oring asks, "What is the source of the authority of tradition and how does its force make itself felt?" (2012, 238).

One part of our answer to these questions should already be obvious: the body itself provides an important source of authority. But the problem is also more complicated than this. We have said that bodies are only human bodies as long as they are involved in social processes, and socialized, enculturated bodies shine a light on yet another core paradox: the universal materiality of human bodies vis-à-vis the singularly unique manifestation of each human's body. Much is at stake when we make the move from the universal to the particular. As neuroscientist Antonio Damasio observes, "The personal and immediate social domain is the one closest to our destiny and the one which involves the greatest uncertainty and complexity" (1994, 169). Generalizable bodily attributes like language organs, bipedalism, or even perception itself tell us little about individual and localized group purpose, about the moment-to-moment social conduits that lead us to our destinies. Performance displays purpose, and a folkloristic explanation of illusions must account for the ways that purpose coincides with perception. We want to show when and how one prompts the other. To accomplish this goal, we will need to take a much closer look at contemporary theories of perception and at how those theories align with the active-transmission theories we detailed in the previous chapter.

Not coincidentally, *activation* and *activity* also feature in what we deem to be the strongest contemporary theories of perception. We will close this chapter by outlining specific systems of active perception in order to help explain the transmission and variability of two folk illusions: one appropriately ancient and the other necessarily social.

Active Perception or the Inner Sherlock

What is active perception? To answer this question, we want to use an analogy. The analogy starts with the premise that humans—even children and youth—are perceptual detectives. We all use hints and clues from current contexts and goals alongside knowledge gained from previous experiences in order to categorize the world around us. To make the analogy clear, it helps to refer to the most famous of detectives, Sherlock Holmes.[1]

In one sense, the long-standing fascination with Sir Arthur Conan Doyle's fictional detective Sherlock Holmes is the antithesis of youth's fascination with illusions. Folk illusions play with fantastic perceptual ambiguity while Sherlock Holmes entertains us with his powers of perceptual acuity, with his ability to detect exact realities. But in another sense, something very *Holmesian* is needed in order for us to manage perception, and that is perceiving via well-formed theories based on current data and past experience. Richard Gregory explains the analogy as a relationship between cues and clues: "Vision works from innate *cues* triggering reflexes and inherited behavior patterns, learned by the genetic code, and in 'higher' animals and man from sophisticated *clues,* based on individual knowledge learned by brains. The ideal example is the inferences of the Victorian fictional detective Sherlock Holmes from small clues, depending on his wide knowledge including brands of tobacco and varieties of human behavior. Although this suggested distinction between cues and clues may not be generally accepted, it seems useful for thinking about origins of perception in evolution and from individual experience" (2012, 176). In reality, we are inundated with cues, with stimuli from our environment. The central thrust of active perception theories is that no stimulus is ever processed outside of governing, clue-based guesses about the nature of the reality from which those cues emanate.

A perfect example of Holmes's well-honed, perceptual prowess shows up in "The Adventure of Silver Blaze" (1892), a short piece detailing Watson and Sherlock's search for a missing prize horse, named Silver Blaze, in the rolling moorlands of southwest England. Silver Blaze has been missing for a couple of days and his trainer, John Straker, has been found apparently murdered in a field close to the stable. We will try not to spoil too much of the story, but suffice it to say that Sherlock finds the horse and gets his man. The case's solution is probably best known for the episode of the dog that does not bark in the night, but we want to zoom in on a separate tiny-but-important clue—a matchstick covered in mud.

Upon their arrival in the moorlands, Sherlock and Watson soon find themselves in the company of the local inspector, Gregory, who has developed a tentative theory about the case. As is typical, Sherlock is not impressed with Inspector Gregory's theories, but more pressing is the fact that the prize horse is still missing just a few days before an important race. Watson, Inspector Gregory, and Sherlock head out to the hollow where the trainer's body was found to look for additional clues. At this point in the story, the crime scene had already been assessed by Inspector Gregory and his men. Moreover, a very heavy rain on the night of the incident had left the ground around the trainer's body in muddy disarray. Regardless, Sherlock desires to scan the spot once more with his own eyes. This excerpt picks up just as Sherlock lies upon the ground to inspect the scene:

> Then stretching himself upon his face and leaning his chin upon his hands he made a careful study of the trampled mud in front of him.
> "Halloa!" said he, suddenly, "what's this?"
> It was a wax vesta [a match], half burned, which was so coated with mud that it looked at first like a little chip of wood.

"I cannot think of how I came to overlook it," said the Inspector [Gregory], with an expression of annoyance.

"It was invisible, buried in the mud. I only saw it because I was looking for it." (Doyle 1976, 180)

Because the muddy match plays an integral role in Sherlock's theory and eventual solution of the case, he is "looking for it" in the mud.[2] Sherlock, here, has brought us face-to-face with yet another enticing paradox: How can anything be at once invisible and visible?

The deceptively simple answer is just this: The match is invisible to those who are not looking for it, but to those who *actively* look for it, it can be—and in Sherlock's case is—seen. It is an answer that helps us to further our analogy, for not unlike Sherlock's match in the mud, we are already expecting (on some level) all of the things that we perceive. When we look upon the bay and see boats on the water, a part of us hypothesized they might be there. When we reach out with a foot in order to locate the gas pedal in our automobile, we do so expecting the gas pedal to be where we find it. When we close our eyes and lean in to our beloved, the lips we feel pressing against our own are felt *as* lips precisely because we are expecting the kiss and all of its loveliness. Let us think of this part of our perceptual systems—the part that is actively deploying theories based upon past experience and current circumstance in order to perceive what it expects to find—as our detective within, as the *inner Sherlock*.

No analogy is perfect, and there are key differences between Sherlock the detective and our analogical inner Sherlock that we need to address. Most importantly, we do not mean to posit a homunculus. There is no little detective buried somewhere in our perceptual systems, at least not in the sense that our inner Sherlock has a consciousness of its own. The homunculus is a well-known regressive fallacy, for if our inner Sherlock were an actual little detective, working from somewhere inside of us consciously controlling our perceptual systems, then our inner Sherlock would need its own inner Sherlock to successfully do its job. We would need to posit inner Sherlocks in perpetuity in order to make sense of it all.[3]

Unlike Sherlock the detective who consciously applies his theory about a missing horse in such a way that he is looking for a match in the mud, we usually *are not* conscious of our inner Sherlock's expectations. In some instances—as might be the case when we rummage at the bottom of a backpack or purse in search of a set of keys—we can be consciously aware of the thing we are expecting to perceive, but this is not necessarily the norm.

We need not walk around consciously processing thoughts like *the floor will be beneath my foot as I step down* or *this cookie will taste sweet when I put it in my mouth*. Nonetheless, expectations correlative to those thoughts certainly do exist, and just like the expert detective Sherlock Holmes who is surprised when he makes an error, we are surprised when our inner Sherlock's expectations are not met. Have you ever stepped on a weak or false spot in the floor only to have your balance sent whirling in confusion? Haven't we all at some point mistaken a salty cracker for a sweet cookie and cringed at the taste of our misplaced expectations? Mismatches like these between our inner Sherlock's expectations and physical reality shine a light on the importance of illusions for the development of contemporary theories of active perception. Like Sherlock the detective, our inner Sherlock can and does make mistakes, and they are mistakes we can learn from.

Active Perception of the Self

Historically speaking, it is appropriate to separate theories of perception into two categories: passive perception and active perception. In the first, perception—sometimes referred to as *sensism* or as *direct perception*—involves the sensory organs' passive and largely accurate presentation of environmental stimuli. In the second, perception—sometimes referred to as *hypothesis-driven* or *indirect perception*—involves active deployment of statistically governed perceptual searches that constantly coalesce external sensations with internal expectations. While passive, direct perception has ancient roots in the teachings of Plato's rivals like Aristippus (c. 435–366 BCE), sensist empiricism is most famously associated with John Locke (AD 1632–1704) and his influential theory of tabula rasa—in which humans are likened to blank slates upon which experience (especially sensory experience) leaves impressions that the mind must then organize, interpret, and assimilate.[4] In theories of passive perception, our world is the world we perceive.

Unlike passive perception, active perception is *always* at least partially uncertain. Proponents of active perception often cite polymath Herman von Helmholtz (1821–1894) and his work on visual perception as the progenitor of contemporary active-perception theories. Studying the physiological structures of the human eye, Helmholtz realized that visual perception requires an interpretation of the curved and inverted sensations impressed upon the eyeballs, which means that the central nervous system must somehow reorient the received visual stimuli to produce the right-side-up, three-dimensional world that we experience. Since visual perception requires processing beyond the recorded sensation, Helmholtz described perceptions as *unconscious inferences*. In this account, what we see as the actual world constitutes an analogy—where the received impressions of light upon the retina are analogous to the content of visual experience. Since analogy assumes a kind of psychological (and usually cognitive) action, Helmholtz's approach grounds perception in the body and in local ecology just as it frees perception from a priori connections to veridicality.

Early in the third volume of his classic *Treatise on Physiological Optics* (1866), Helmholtz further articulates his notion of unconscious inferences beyond just visual perception with an early reference to what is now known as phantom-limb illusions:

> For example, the stimulation of the tactile nerves in the enormous majority of cases is the result of influences that affect the terminal extensions of these nerves in the surface of the skin. It is only under exceptional circumstances that the nerve-stems can be stimulated by more powerful agencies. In accordance with the above rule, therefore, all stimulations of cutaneous nerves, even when they affect the stem or the nerve-centre itself, are perceived as occurring in the corresponding peripheral surface of the skin. The most remarkable and astonishing cases of illusions of this sort are those in which the peripheral area of this particular portion of the skin is actually no longer in existence, as, for example, in case of a person whose leg has been amputated. For a long time after the operation the patient frequently imagines he has vivid sensations in the foot that has been severed. He feels exactly the places that ache on one toe or the other. Of course, in a case of this sort the stimulation can affect only what is left of the stem of the nerve whose fibres formerly terminated in the amputated toes. Usually, it is the end of the nerve in the scar that is stimulated by external pressure or by contraction of the scar tissue. Sometimes at night the sensations in the missing extremity get to be so vivid that the patient has to feel the place to be sure that his limb is actually gone. ([1910] 2005, 3–4)

Like Melville's Captain Ahab, who begs the *Pequod*'s carpenter to make him a wooden peg that will forever drive away the "old Adam" of his lost leg, the patient who grasps for a lost appendage in the night cannot overcome the lingering perception of the limb that was there (Melville [1851] 1992, 481).[5] For Helmholtz, the now-missing leg remains in the very present, very active perceptual inferences that continue to presume wholeness.

In recent decades, phantom perceptions have taken center stage on several issues of mind and body like pain, neuroplasticity, and body image. Probably the best-known examples of this research come from V.

S. Ramachandran and his colleagues who created a system of treatment for phantom-limb pains involving mirror visual feedback:

> There is a sense in which one's body image is itself a "phantom": one that the brain constructs for utility and convenience. A striking illustration comes from studies of amputees who experience phantom limbs. We had patients insert their "good arm" and phantom arm through two holes in the front of a "virtual reality box." The roof of the box was removed and inside a vertical mirror divided the holes in the sagittal plane. The patients viewed the reflection of their intact hand in the mirror, thus creating the illusion of observing two hands. Several subjects viewing the reflection while the intact arm was touched reported feeling the touch on the phantom limb. (Armel and Ramachandran 2003, 1499)[6]

We will have much more to say about mirrors in chapter 6, but it is important to note here that Ramachandran's work with amputees' distorted self-perceptions demonstrates that body images are malleable, that they "can be profoundly altered by the stimulus contingencies and correlations that one encounters" (Armel and Ramachandran 2003, 1506). In the contexts of Helmholtz's unconscious inferences, pathological phantom pains demonstrate the always ongoing construction of our bodies.

Another component of Ramachandran's research program—a component more relevant to our work with folk illusions—deals with experimental contexts that give rise to phantom perceptions in healthy subjects (i.e., subjects who do not suffer from pathological phantom pains as a result of amputation or nerve damage). In these studies, haptic perception is reportedly projected from the body of the subject into an external object such as a rubber hand or even the top of a table:

> The subject sits next to [a rubber hand] and places his own corresponding hand (say his right hand) next to it. A vertical partition is then placed on the table in between the real and fake hands so that the subject's view of his real hand is occluded but he can see the fake hand. While the subject looks at the fake hand, the experimenter applies a long sequence of randomly placed strokes and taps on it while at the same time the hidden real hand is stroked and tapped in synchrony. The subject then experiences the uncanny illusion that the touch sensations are actually felt in the spatial location of the dummy hand—not from the hidden real hand! . . . The second effect, which was observed in our laboratory, is even more surprising. The subject's real hand is hidden by a partition as in the previous experiment. However, instead of using a dummy hand, we simply stroked and tapped the table in precise synchrony for about a minute. To our astonishment, subjects often reported sensations arising from the table surface, despite the fact that it bears no visual resemblance to a hand. (Armel and Ramachandran 2003, 1499)

For us, Ramachandran's projected-perception experiments provide another example of an interactive context within which people can manipulate each other's perceptions. Like a magician, the experimenter casts the subject's sense of self into a world that only runs parallel to normal perception, and in this parallel world, the scientist's experiment shares an important function with children's performances of folk illusions—the co-activation of bodily quirks.

Ramachandran, Shakespeare, and Intersubjectivity

Claiborne Rice

The similarities between the participant roles featured in the rubber-hand experiments and the agent/patient roles we have described in several folk illusions are undeniable. Still, scientists do not seem to be aware of (or simply do not care about) folkloric performances that play with the malleability of embodied perception. Ramachandran and his colleagues open a 2011 discussion of experiments that involve the projection of phantom perceptions onto the head of a mannequin with the now-familiar scientistic allusion to the nonscientist's naivete: "You take your body—that 'muddy vesture of decay' surrounding you—for granted. You do not doubt that your body is your own, or attribute your sensations to other people. Yet there are many clinical cases and artificially contrived laboratory situations in which this assumption is called into question and your body image is profoundly disrupted" (Ramachandran, Krause, and Case 2011, 367).

For a scientist who accredited his son with teaching him Floating Arms (see chap. 2), it is strange Ramachandran does not recognize, here, the widespread youthful play associated with bodily illusions. The reference from Shakespeare's *Merchant of Venice* that he uses to authorize his point conjures up the Platonic notion of appearances diminishing the real that Shakespeare inherits from the Renaissance. In the play, Lorenzo is flirting a bit with Jessica in the evening on the banks of a stream:

> How sweet the moonlight sleeps upon this bank!
> Here will we sit and let the sounds of music
> Creep in our ears. Soft stillness and the night

Become the touches of sweet harmony.
Sit, Jessica. Look how the floor of heaven
Is thick inlaid with patens of bright gold.
There's not the smallest orb which thou behold'st
But in his motion like an angel sings,
Still choiring to the young-eyed cherubins.
Such harmony is in immortal souls,
But whilst this muddy vesture of decay
Doth grossly close it in, we cannot hear it. (5.1.54–65)

So, yes, for Shakespeare, a child of the Renaissance, the body trapped in the world of appearances does impede one from recognizing a higher order clarity or beauty. But all is not lost! Such attention to beauty can be learned. As the scene goes on, musicians come in, and Jessica remarks, "I am never merry when I hear sweet music." Lorenzo replies to her that "her spirits are attentive," but

The man that hath no music in himself,
Nor is not moved with concord of sweet sounds,
Is fit for treasons, stratagems, and spoils.
The motions of his spirit are dull as night,
And his affections dark as Erebus.
Let no such man be trusted. Mark the music. (5.1.82–88)

Mark the music. If you attend to music, that image of Platonic ideal perfection, you can learn to recognize the ideal beauty and pattern in the world. You will defend your spirit from "treasons, stratagems, and spoils." Though our vesture is muddy, we can still leverage the bits of beauty that we can perceive to help us turn our affections from the dark into the light, to see truth and beauty more clearly. If we take our body for granted, as Ramachandran says we do, we are not marking the music. Is he suggesting that we all ought to decamp to the neuroscience lab to get a reminder of how fallible and imperfect our bodies are? Or take in an opera? Instead, what if we were to attend to the music of our inner spheres, observe the motions of our affections as our body takes up strange or unfamiliar positions? As we thematize our own body in the act of perception, as we make our own body the focus of perception, we can perhaps make our body visible again.

When viewed as a genre, folk illusions appear to serve this purpose. If we want to be Platonic, we can say that illusions point us toward the realization that our systems of perception are imperfect, incomplete, or restricted. The shortcomings of the systems advertise the body as a contingent system rather than allowing us to take the world simply as it comes. They teach us to question that system and seek a more abstract understanding of the real.

Come to think of it, Shakespeare is working in a dramatic mode, which itself takes for granted a certain quality of intersubjectivity, of knowing that others share our responses to a dramatic situation. Lorenzo is instructing Jessica on the nature of music and human perception, something that also assumes a certain amount of intersubjectivity. Their whole discussion is rich with assumptions regarding the sharing or overlapping of experience, in much the same way that folk illusions depend on and reinforce these assumptions. What would be the purpose of trying to communicate our thoughts and feelings if we weren't in some sense taking for granted a degree of similarity between our various bodies and minds? If folk illusions help us to understand the contingent nature of everyday perception, then they also teach us that on the whole, as long as we mark the music, we will be fine.

Taken together, Helmholtz's reference to the amputee's unconscious inferences and Ramachandran's work with mirrors and rubber hands more than a century later exemplify a dominant movement toward theories of active perception in which statistical amalgamations of past experience govern the likelihood of any given perceptual outcome. Here is Armel and Ramachandran's explanation of the principles that govern the illusory phenomenon in the rubber-hand and table-top experiments: "In addition to demonstrating the malleability of body image, this simple illusion also illustrates an important principle underlying perception: that the mechanisms of perception are mainly involved in extracting statistical correlations from the world to create a model that is temporarily useful" (Armel and Ramachandran 2003, 1506).

The scientists further describe the principle as a kind of perceptual binding that occurs across separate modalities: "We suggest that the principle underlying this illusion is Bayesian [statistical] perceptual learning—that two perceptions from different modalities are 'bound' when they co-occur with a high probability. In the hand illusion, the seen and felt touch were bound because of their temporal synchrony" (Armel and Ramachandran 2003, 1505).[7]

Evidently, the tendency is so strong that even the surreality of *having a table top for a hand* cannot sever the perceptual bind: "In this case, the brain's tendency [is] to take advantage of statistical correlations (even when they do not 'make sense' from the cognitive point of view and contradict a lifetime of experience with our own bodies). Thus, instead of emphasizing the visual resemblance of the fake hand to the real hand, we would place emphasis primarily on the synchronicity of stimulation" (Armel and Ramachandran 2003, 1505).

Perception, then, reaches out into the world probing for likely ecologies, intermingling vestiges of past experience with present circumstance. In this line of thinking, neither veridical reality nor the mechanisms of perception preempt any given phenomenological experience. Only the intermingling is real. This is active perception's most condemning critique against sensism.

An Ancient Folk Illusion?

Can active perception provide insight into the nature of cultural constancy and differentiation? In 2009, we organized our first observation of four eleven- and twelve-year-old girls in South Louisiana, and there we saw for the first time a performance of what might be thought of as the oldest folk illusion.

After everyone gathered at the home of twelve-year-old Monica, Monica's mother and older sister watched as we prompted the girls to perform several forms like Floating Arms and Twisted Hands in the backyard. Having quickly understood the kinds of activities we were interested in, Monica demonstrated an elegant, self-taught illusion. First, she crossed her middle finger over her pointer finger on both hands. Then, she touched the tips of the four, crossed fingers. Without looking and while maintaining contact between the crossed fingertips, Monica twisted her forearms and wrists. When asked why she liked performing this activity, Monica simply replied that doing so made it difficult "to *feel* which tips of which fingers are touching which tips of which fingers." Unbeknownst to her, Monica had just (re)created a variant of the ancient Aristotle's illusion.

Aristotle's illusion shows up frequently in western writing. Many classicists, psychologists, and experimental scientists have reported that the first description of the illusion appeared in Aristotle's philosophical works.

Fig. 4.1. Crossed Fingers, known traditionally as Aristotle's illusion, is typically performed at the nose.

In *On Dreams* (350 BCE), the Greek philosopher alluded to the experience of touching a single object with crossed fingers just as he highlighted the importance of multisensory inputs for judging reality to be real: "The reason why these things happen is that the ruling part and that by which appearances occur do not judge on the basis of the same faculty. Proof of this is the fact that the sun appears only one foot across, and yet frequently something else contradicts the appearance. Again, by crossing of the fingers a single object appears as two, but even so we still deny that there are two things. For sight has more authority than touch. If touch stood alone, we should actually judge the single object to be two" (quoted in Gallop 1990, 93).

Aristotle also referenced the crossed-finger illusion in Book IV of his *Metaphysics* (350 BCE), where he describes a relativistic approach to truth:

> For to those who for the reasons named some time ago say that what appears is true, and therefore that all things are alike false and true, for things do not appear either the same to all men or always the same to the same man, but often have contrary appearances at the same time (for touch says there are two objects when we cross our fingers, while sight says there is one)—to these we shall say "yes, but not to the same sense and in the same part of it and under the same conditions and at the same time," so that what appears will be with these qualifications true. (Ross 1924)

Let us leave aside, for the time being, multisensory perception and the nature of truth to focus on what we might make of twelve-year-old Monica recreating the famous illusion in the twenty-first century. Monica, as she explained, just happened upon her version of the activity by chance, by playfully fiddling with her fingers as children so often do. Considering our central argument that people—even children—are highly attuned to the illusory tendencies of perception, you may no longer be impressed by a twelve-year-old stumbling upon a version of Aristotle's illusion, but taken together, the parallels between Monica's and Aristotle's cross-finger illusions alongside the parallels between Monica's and Aristotle's respective analyses of the illusions point toward a kind of stability inherent in the genre.

Aristotle makes no mention of the pleasure derived from the illusion, but otherwise his observations align quite well with Monica's: Crossing our fingers leads to ambiguous haptic perceptions; the connections between veridical reality and perception are tenuous.[8] Systematic study of crossed-finger illusions à la Aristotle's illusion has been a mainstay of experimental science since the middle of the nineteenth century when Johann Czermak reanalyzed the "famous trial" of crossing one's fingers in 1855; specifically, Czermak demonstrated that crossing one's fingers while touching an object with a rounded surface on one side and a pointed surface on the other inverts the perception of two distinct surfaces. He also showed that a similarly ambiguous experience can be created by distorting one's lips. When our mouth closes normally, each lip touches the corresponding point on its opposite along its whole continuum. An object touching each lip at the correlative points is sensed as a single object. Place a (clean) single object like a pencil or your finger between your lips and you will feel one pencil. Then try distorting your lower jaw to one side (the farther you distort it, the better the illusion will work) and closing your mouth so that the

Fig. 4.2. Monica (now age twenty) demonstrates her version of a Double Crossed Fingers illusion.

normally correlative points along the lips are out of synch. Take the same pencil and put it in your mouth, and you will distinctly feel two separate pencils! Looking down at the pencil or looking at yourself in the mirror while doing this will not diminish the illusion one bit. Since Czermak's discovery, others have reported that illusions of displaced and/or ambiguous crossed-tactile perceptions can be created on the face, ears, and scrotum (see Benedetti 1985, 518).

Like scientists who create variants in their labs, children and youths who play with crossed-finger illusions both recreate and extend ancient philosophical insight. Monica's self-taught version isn't the only remembrance like this we have gathered. In the spring of 2015, college freshman Kaydee Akins, who was raised in Columbus, Indiana, reported learning the activity from a playmate in elementary school: "One [illusion] I remember growing up involved crossing your middle finger over the top of your index finger. You'd then use your opposite hand's pointer finger and touch the spot where the two fingers were crossed, and it'd feel as though two fingers were touching that place."

Contemporary folkloristics rarely concerns itself with origins, and folklorists—working, for example, with verbal and musical traditions—are well accustomed to a particular saying or song making its way from the more authoritative traditions of writing into the anonymously authored complex of face-to-face, oral communication. Orally transmitted variants of Aristotle's illusion reiterate the importance of immediacy in folkloric performance because, as we mentioned, none of the performances we have observed or remembrances we have collected have been characterized by performers as an extension of Aristotle's illusion.

Youthful performers with whom we have talked are also unaware of the illusion's scientific history. Neither the philosophical nor the experimental history is traced in the children's traditions; the illusion and the awareness of the illusion's attendant perceptual processes remain anchored in the polygenesis of self-awareness *and* in the locally manifest communal awareness that follows. For these reasons, we differentiate the folk illusion from the scientific study of the illusion by calling the children's versions Crossed Fingers (see A1, A1a, and A1b). That being said, we must admit once again that we have no reason to rule out the possibility that Aristotle himself was commenting on a form of living folklore when he referred to crossed-finger illusions some 2,300 years ago. We must leave open the possibility that crossed-finger illusions have always shown up as a part of children's play and that they have maintained both a folkloric and a critical heritage since—and possibly before—Aristotle.

Our Inner Sherlock as an Explanation for Crossed Fingers

Richard L. Gregory, a self-professed intellectual descendant of Helmholtz and the twentieth-century champion of active perception, famously took some illusions to be special cases of perception that expose the hypothesis-driven nature of active perception. He often compared the hypotheses of active perception to the hypotheses of science, arguing that like the hypotheses of science, the hypotheses of perception can fail—misleading perceivers and scientists alike. Gregory's analysis of Aristotle's illusion neatly summarizes his illusions-as-misplaced-hypotheses theory: "An example is Aristotle's illusion. Try crossing the first and second fingers of one hand. If the nose is rubbed gently with the now inner surfaces of the crossed fingers—which are normally their outer edges—one may experience, by touch, two noses! This is clearly due to an incorrect assumption about the positions of the fingers, and so the regions in contact with the nose" (1973, 70). For Gregory, the "incorrect assumption" is not a priori. Experience, as a matter of fact, is key. The incorrect assumption arises from the vast majority of tactile experiences in which the fingers are not crossed.

Gregory also describes perceptual hypotheses as applied knowledge, where "knowledge is information structured for use," and "perception depends on knowledge" (2009, 111), which is to say knowledge of past experience. Here, we have exposed the methods of our inner Sherlock: correlating experiential tendencies with active perceptual probes. In the case of Crossed Fingers (a.k.a. Aristotle's illusion), the illusory perception of a single pointer finger with two tips or of two noses depends on the experientially derived "knowledge" of touching objects with the usually uncrossed pointer and middle finger of a single hand.

We can get a feel for our inner Sherlock's hypotheses with the simple exercise of rubbing the outside of one hand's *uncrossed* middle and index finger. Begin by pointing the two fingers outward; then take one finger from the other hand and rub the outside of the index finger (the side closest to the thumb), then the outside of the middle finger (the side closest to the ring finger). Repeat these steps a few times. Now, leaving the fingers uncrossed, take two fingers from the opposing hand and rub the outside of both the middle and index finger simultaneously. How much experience supports the hypothesis that an object (or objects) simultaneously touching both the outside of the middle finger *and* the outside of the index finger *must* be separated by at least the distance equal to the sum width of those fingers? For those of us who experience crossed-finger illusions, the power of this hypothesis is so great that our inner Sherlock prefers the reality of having two noses over some other reality that confounds the past experience of the fingers.

Now, active perception's importance for folk illusions begins to come into focus: (1) Hands reach beyond the individual, and active perception transcends the individual. (2) The Crossed Fingers illusion exemplifies the products of active perception. (3) The physiological and psychosomatic processes of the hand constitute an important source of authority for traditional play with the illusion. (4) The exploration that accompanies a performance of crossed-finger illusions—whether social and traditional or solo and spontaneous—exemplifies the kind of self-reflection that leads to self-awareness. These insights help us to understand how an otherwise trivial activity might persist.

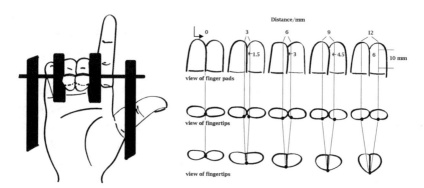

Fig. 4.3. *Left:* Benedetti's hand clamp. *Right:* As the clamp is tightened, the middle- and ring-finger pads are increasingly displaced. When the finger pads were displaced by the largest increments of nine to twelve millimeters, subjects reported an illusory double-tactile sensation in 85 percent to 96 percent of trials. (Recreated with permission of the author.)

Experimental evidence confirms and extends Gregory's explanation of crossed-finger illusions, which means that we can use variants created in the scientists' labs as a kind of gloss on variants found in children's play. Frequently cited is the work of physiologist Fabrizio Benedetti, who published a series of studies on the phenomenon in the last decades of the twentieth century. In 1986, Benedetti named the experience of illusory double-tactile sensations *diplesthesia* after the visual phenomena *diplopia*, commonly known as double vision. Like the double tactile stimulations of crossed-finger illusions, many of us can voluntarily give rise to double-visual sensations simply by focusing our sight upon a single object, a can of soda for example, and crossing our eyes. Similarly, Benedetti's work with diplesthesia highlights the shared features between crossing one's fingers and crossing one's eyes by focusing on the actions that lead to irregular finger positions. More drastic actions lead to less likely finger positions, and these less likely finger positions are more likely to give rise to illusory experience.

In one specific experiment, Benedetti showed that subjects whose ring and middle fingers are placed in a specialized hand clamp that pressed together the normally separated areas of the two finger pads (the part of the fingertip commonly copied for fingerprint records), report illusory double-tactile perceptions when a single object is pressed against those abnormally adjacent skin points (see fig. 4.3). By tightening and loosening the hand clamp, Benedetti could control the subjects' fingers to create more or less likely finger positions. He discovered a strong correlation between the amount of pressure applied by the hand clamp (i.e., the less standard finger-pad position) and the tendency to report the illusion.[9] If you remember playing with vision by crossing your eyes (and we expect many of us have such memories), then you will quickly recognize the continuing parallel between diplopia and diplesthesia—the more crossed your eyes become, the more doubled your vision.

It is in the midst of all this doubling that Monica's variation of Crossed Fingers takes center stage. Her variant is, of course, a double-crossed-finger illusion. Ask yourself, how many times have youth's hands and fingers touched one another by the age of twelve? And subsequently, how many times do any of us touch the tips of our pointer and middle fingers against one another while in a double-crossed-finger position? Without conscious awareness of the statistically governed quality of active perceptual hypotheses, Monica's double-crossed-finger performance position serves the same purpose as Benedetti's hand clamp, for both create a bodily posture that is even more statistically unlikely than a prototypical crossed-finger position. Here, we can identify one way that active perception facilitates folkloric variation: Crossed-finger illusions—taken as a characteristic illusion of hypothesis-driven, active perception—occur when finger posture and statistically governed assumptions about finger posture misalign during tactile stimulation. That being said, performances of Crossed Fingers can vary in accordance with the methods of our inner Sherlock by assuming performance positions that are increasingly less likely.

Active Perception and the Oldest Trick in the Book

We have met our inner Sherlock, recognized it as a characteristic of our perceptual processes, and identified its methods. We have learned how those methods (i.e., active, hypothesis-driven perception) give rise to an ancient folk illusion, Crossed Fingers, *and* how those methods allow for variation of the illusion. But before we conclude our discussion of active perception, we want to apply these lessons to a highly social context: the primary contexts for folkloric performance, transmission, and variation. We do so by introducing a heretofore-unmentioned folk illusion, the sheer familiarity of which seems to have rendered it utterly invisible as a folkloric performance. We like to call it Who's Touching You?

Because of its pervasiveness, Who's Touching You? has a strong presence in both popular and folk culture. One of our favorite descriptions of the tradition—and one that we feel fully obliged to share—can be found in the satirical online encyclopedia, *Uncyclopedia: The Content-Free Encyclopedia*, on the page for "The Oldest Trick in the Book": "The Oldest Trick in the Book is the infamous 'Tapping on a person's left shoulder when you're standing on their right.' This trick was first chronicled in cuneiform by the Ancient Sumerians. This chronicalisation also created 'The Book' itself. In this article, we will chronologically summarize, from oldest to newest, the tricks in The Book."[10] Of course, you recognize this trick. We are (almost) certain you have witnessed and experienced it. But when was the first time you experienced or witnessed it? When was the last time? How many times have you performed as the agent? How many times as the patient? Where could such an endemic folk illusion have come from?

Like Crossed Fingers's historical relationship to Aristotle's illusion, Who's Touching You?'s historical relationship to *the oldest trick in the book* remains entirely unclear. The opportunities for polygenesis and diffusion are both so great that we can never expect to have a full history. That being said, *Uncyclopedia* (with a respectable amount of humorous grace) offers the following:

The Shoulder Tap

In the year 10580 B.C.E. the Babylonian king, Hammurabi, reigned supreme over the Mesopotamian deserts. On February 30, during a long speech by Irhemhotep, he stood on Shamadad's right, and reached over and tapped him on the left shoulder. As a result, Shamadad looked to his left, where no one was standing. All would laugh, as decreed by law, or else be condemned to read Oscar Wilde quotes everywhere. Immensely proud of his ingenuity, Hammurabi ordered these endeavors to be recorded in cuneiform on a clay tablet. This tablet was the first page in a small "tricks" archive, which would later become "The Book."

Alternative Shoulder Tap

Years later, the minor prophet Zarathustra, experimenting with various tricks, attempted what was previously unheard of: the other side. Standing on the *left* this time, and tapping the person's *right* shoulder was, until that time, never considered, and almost heretical. It worked with stupendous results, even on people who have been so duped many times before.[11]

In its satirical success, *Uncyclopedia*'s description of the Shoulder Tap contains important nuggets of truth. The trick is powerful. The trick works from both sides. It can be performed successfully in uncountable contexts, and a single patient can be duped an untold amount of times. The intended illusion in these tapping tricks is inherently social, which is the main reason why we choose to name the activity Who's Touching You? as opposed to less objective names like *shoulder tap* and *alternative shoulder tap*.

In line with the genre of folk illusions, Who's Touching You? involves previously discussed participant roles of agent and patient. A patient turning to the side of her touched shoulder is befuddled by the absence of the presumed agent . . . who stands laughing on the other, nontapped side. The agent's laughter or, in some variants, her intense focus on something—anything—else signifies the playful quality of the tradition. The illusion works by way of misdirection, and its ubiquity demonstrates shared cultural knowledge about reflexes and attention. Considerable effort in the sciences of mind has been devoted to studying the nature of attention more generally. Starting with the work of Michael Posner in the 1970s, experimental scientists have spent much time investigating the neurological and cognitive aspects of attention—especially in regards to the spatial orientation of attention, using the *spatial cueing paradigm*.[12] These scholars draw a distinct line between *exogenous* and *endogenous attention*—where exogenous attention refers to conscious attention that is a reflex or reflex-like reaction to external, perceptual stimuli, like being tapped on the shoulder, and where endogenous attention refers to conscious attention that is voluntarily focused on some internal (e.g., attending to the way your right foot is feeling right now) or external (e.g., attending to a spot on the ceiling above your head) entity. Another important distinction is that of *overt* and *covert* attention. Overt attention coincides with observable eye, head, and other postural movements that orient our bodies in the direction of a target, while covert attention involves entirely subjective shifts in attention that cannot be observed by onlookers. In this system of categorization, we would say that a prototypical performance of Who's Touching You? plays with the patient's overt attention to an exogenous stimulus resulting from interpersonal touch.

Of course, attention serves an important role in active perception. Remember that Sherlock the detective's attention to the minute contours in the mud leads him to finding the otherwise invisible matchstick. Richard Gregory admits that our vulnerability to being tricked by others often grows out of the nature of active perception: "This reliance on hypotheses of what is out there beyond what is signaled makes conjuring relatively easy. Conjuring shows how fragile perception really is and how far it depends on assuming normal objects doing their usual things" (2009, 107).[13] More-over, unexpected touch demands our attention because unlike seeing or hearing—in which stimuli can reach our perceptual organs from a great distance—anything that touches our body is right here, right now. We are reminded, once again, of the fearful attention that accompanies a tickling sensation that may or may not be a spider. Can you *not* attend to the feeling of little legs crawling on the back of your neck, whether or not that feeling coincides with the presence of an actual creepy crawly?

Interestingly, Who's Touching You?'s touch is more benevolent than the malicious insect. The shoulder tap is marked as interpersonal touch just as it is being perceived, for what else in the world taps our shoulders in such a way as to gently gain our attention? The perception of the tap is never *only* the perception of something touching our back, it is always the perception of the genuine person who our inner Sherlock assumes is standing on the tapped side—a hypothesis about the world beyond the signal.

Active Perception in a Crowd, or Is Sammich Touching You?

Claiborne Rice

Like our analysis of Ping-Pong in the previous chapter, our under-standing of Who's Touching You? was significantly shaped by a fortuitous observation of an unprompted, natural performance of the tradition. On April 18, 2016, my friend, Charles "Big Sammich" Caldwell, and I went to see the free Lil Wayne concert at the Cajun-dome on the University of Louisiana at Lafayette campus. UL's chap-ter of Alpha Phi Alpha fraternity had collected more than seven thousand bottles of water for the town of Flint, Michigan in response to the "Social Wave for Change" challenge from Wayne's music com-pany, Tidal.[14] Tickets were free for all UL students, faculty, and staff. I invited Sammich, as he had been a student on and off for ten years but did not currently have an ID.

We showed up a half-hour before the doors opened. The crowd waiting to enter was large, and there were only four doors opened, with security searching and wanding everyone who entered. The lines moved slowly, and a little rain was falling, so people were pressing

Fig. 4.4. Charles "Big Sammich" Caldwell with his friend Clai Rice.

in under the entrance portico as they waited. The crowd was over-whelmingly college age, and races were well mixed. It got tighter as we moved closer to the door. I was glad to be with Sammich, a large, cheerful, African American gentleman who is about six-foot-four and closing in on three hundred pounds and so is easy to spot in a crowd.

At a point about fifty feet from the door, a group of four or five friends—white, college-age men—had moved up on our left beside us in the crowd. One young man, about five foot eight with darker blonde hair, was standing just to the left and slightly behind his friend, about six feet with black hair. The taller guy was turned away from us a bit, half-facing his friend and looking mostly over his shoulder into the crowd as the two talked intermittently. As they shuffled forward slightly in the crowd, the shorter fellow reached his hand out behind his friend's back and touched him lightly on the right side of his neck. The friend did not react.

They started talking, and as they talked, the guy reached out and touched him on the neck again. The taller guy did not react immediately, but in the next few moments he shuffled around—moving in the crowd was difficult—to face forward toward their other friends, which enabled him surreptitiously to get a view of who was standing in the crowd to his right, the direction that the pokes to his neck ostensibly had been coming from. He clearly saw Sammich standing there and quickly turned back around to converse again with the shorter fellow. As they talked, the shorter guy reached out yet a third time and touched him on the neck. If the taller guy had seen or suspected his friend, I would expect him to have jokingly confronted him. To his credit, though, he neither confronted his friend nor opened a dialog with Sammich. Eventually they moved on ahead in the crowd.

Like the stolen-ball variant in a performance of Ping-Pong, this performance (which we like to call Is Sammich Touching You?) actually presents a truncated version of the tradition, because the patient never actually turned toward Sammich's direction. But we can learn a great deal by asking ourselves why the patient was not tricked, why the folk illusion failed.[15] To begin, it is important that the agent did not try to draw the patient's exogenous attention via the prototypical shoulder tap; instead, he lightly touched the patient's neck—a more intrusive interpersonal exchange. The agent was clearly trying to effect a *who-the-hell-is-tickling-my-neck* response so that the patient would turn, in accusation, toward a hulking Sammich. In a crowd, we can anticipate certain types and body locations for contact, but the neck is unlikely to be touched accidentally. In the contexts of active perception, we may hypothesize that the haptic sensation was so unlikely the kind of interpersonal touch that would require turning toward the touch's direction that the patient actively perceived the tickle as a trick.

Recognizing that attention depends a great deal on situational contexts and current tasks, we might similarly hypothesize that the patient had already noticed Sammich—who is easy to notice—and was reluctant to turn that way even before his neck was tickled. Possibly, past experience with the agent had piqued the patient's attention in such a way that he was already alert to the likelihood of a trick. We remain agnostic as to the precise reason for the flawed performance, but what is most important about these possible explanations is the fact that they fully embed contemporary theories of active perception in social contexts that foreground both bodily experience *and* folkloric tradition. It is this very combination, after all, that leads us to categorize the oldest trick in the book as a folk illusion in the first place.

Active Perception and Lessons from the Field

Categorizing Who's Touching You? as a folk illusion is not a trivial matter. For one thing, we know of no other work that treats the traditional trick as an illusion, so it is a categorization that allows us once again to assert folk illusions' importance to the study of illusions more generally. It is an assertion that is paralleled by neuroscientists Stephen L. Macknik and Susana Martinez-Conde's nearly folkloristic work with magicians. Back in chapter 1, we briefly referred to Macknik and Martinez-Conde's excellent book, *Sleights of Mind* (2010), as an example of a recent influx of scientific attention that is being given to stage magic.

Sleights of Mind presents an engaging, accessible version of Macknik and Martinez-Conde's technical argument that experimental neuroscientists can and should examine and learn from magicians' artistic craft. Their desire to learn about the magicians' insights into the nature of performing illusions closely parallels strong and honest ethnography: "Magicians have tapped in to the power of cognitive illusions more effectively than scientists have, though less systematically. The magician's goal is to misdirect you and create a sense of wonder. . . . Our goal is to take magic into the neuroscience laboratory and to use it to . . . increase the rate of discovery about our cognitive processes. We believe that magical methods will prove invaluable in determining the circuits in the brain that process cognition, as well as in revealing important new perspectives on how the brain functions" (2010, 252). From the folklorist's point of view, the "magical methods" referred to in this passage represent the traditionalized knowledge and practices shared by professional magicians. Put differently, magical methods are lessons from the field—problematizing data sets—that the scientists use to strengthen their experimental designs.

Macknik and Martinez-Conde provide an interesting example of an applicable lesson from the field: magicians' folk knowledge about curved versus straight motions as misdirecting actions. Magicians' actions on stage are rarely trivial or pointless, and even seemingly innocent gestures or postural changes can be used by the magician as actions for (mis)directing their audience from the "trick" of the trick, such as the palmed coin, fake fingertip, or secret pocket. It turns out that some magicians believe that curved motions are more effective in misdirecting the audience's attention than straight actions. Because the exact relationship between curved-versus-straight motion and attention in neuroscience remains an open question, the lesson is a good example of the kind of traditionalized knowledge that neuroscientists can learn from magicians and test in the lab.

Can folk illusions, like the folklore of magicians, provide applicable lessons from the field for scientists? We think so. A good, recent example is Peter Brugger and Rebekka Meier's "A New Illusion at your Elbow" (2015), which appeared in the scientific journal *Perception*. Brugger and Meier's study is rare in that the scientists mention the exact children's folkloric tradition from which they learned their lesson:

> Among the manifold illusions of the cutaneous sense, some are not readily amenable to playful experimentation as they require substantial technical equipment. One exception is a game Swiss children typically enjoy in playgrounds. They stimulate the inner side of a friend's forearm by slowly moving a finger from the wrist towards the crook of the elbow. Eyes closed, the friend has to shout "stop!" on feeling the crook being reached. On opening the eyes, there is much amazement about an anticipation error, frequently in the order of several centimeters. We investigated the crook-of-the-elbow illusion under controlled conditions and speculate about its potential origin on psychological and physiological levels. (219)

Apparently, the tradition is cross-cultural. In 2009, we fieldworked a performance of what Brugger and Meier call a crook-of-the-elbow illusion in South Louisiana. We also cataloged the play form as the folk illusion Where Am I Touching You? in our initial *Journal of American Folklore* article in 2012:

> Where am I Touching You?—The actor closes her eyes and stretches out one of her arms, palm up. The director pinches the actor's forearm near the elbow crease. Then, with the actor's eyes remaining closed, the director slowly and gently runs the tip of her finger in a circuitous path from the actor's wrist up the arm toward the previously pinched spot. The actor must state when the director has arrived at the pinched spot. Invariably the actor calls arrival well before the director has actually arrived at the spot. When the actor opens her eyes, she is mystified why she felt that the spot had been reached (observation of E. Hesse, age 23, and E. Tomingas, age 25, Spring 2009). (464)

Under "controlled conditions," Brugger and Meier confirmed what the children already knew about the effectiveness of the illusion: "Our measurements confirm a powerful and robust illusory anticipation of touch at the elbow crook when the tactile stimulus is slowly moved in a proximal direction starting at the wrist" (2015, 219). Interestingly, they also gained technical insights as they formed a new hypothesis about haptic illusions and active, hypothesis-driven perception.

Brugger and Meier explain that previous experimental studies with haptic illusions have frequently worked with abnormally fast movements that give rise to illusions. The most famous of these experiments deal with the Cutaneous Rabbit illusion, in which taps or light touches delivered on separate regions of the skin are perceived as an illusory series of linear taps. For example, Frank Geldard and Carl Sherrick, who first identified the Cutaneous Rabbit forty years ago, found that mechanical pulses rapidly delivered at three points on the under forearm starting at the wrist and ending at the elbow are consistently perceived as several small taps as if a small rabbit were hopping along the subject's forearm. In the framework of statistically governed active perception, several theorists have argued that the Cutaneous Rabbit illusion arises from the speed with which a subject's skin is touched in the experiment. "Under natural circumstances," objects touch the skin much more slowly than the touches delivered during the Cutaneous Rabbit studies, so the experimental subjects' inner Sherlocks mistakenly assume that the taps are delivered from a slower stimulus, hopping all along the forearm path. Or, as the scientists put it, the skin regions actually touched by the stimulus during the Cutaneous Rabbit experiments are contracted, eliminating the nontouched space between.

Of course, children's play with Where Am I Touching You? depends on the opposite scenario, which is to say that the folk illusion studied by Brugger and Meier arises from children's misperception of the relatively slow movement of the haptic stimulus. As we wrote in 2012, the agent "*slowly* and gently runs the tip of her finger in a circuitous path from the actor's wrist up the arm toward the previously pinched spot" (Barker and Rice 2012, 464). How many haptic stimuli in "natural circumstances" move as slowly *and* circuitously as the agent's fingertip? Having learned a lesson from the children's playgrounds in Switzerland, Brugger and Meier add a new hypothesis to the science of haptic illusions: "The cutaneous motion illusion studied here occurs in response to a stimulation velocity at or even below the velocities typically experienced in everyday life (e.g. during caressing movements or the crawling of an insect). Hence, if violated at all, a participant's expectations of the speed of a tactile motion would lead them to experience an *enlarged* track on the skin" (2015, 219). If haptic perceptions that are so fast that they trick people's inner Sherlock cause a contraction of the perceived area of the touched skin, then haptic perceptions that are so slow that they trick children's inner Sherlock cause an enlargement of the perceived area of the touched skin. The enlarged area, their theory holds, easily subsumes the space in which the patient in a performance of Where Am I Touching You? reports to the agent that she has reached the crook or pinched spot on the elbow. Truly, it is a compelling lesson from the field.

In this light of the scientists' demonstration that lessons learned from Where Am I Touching You? can be beneficial to the study of active perception, let us close this chapter by reconsidering Who's Touching You? in the hopes of identifying a parallel. As we have already stated, misdirection plays a key role in a performance of the illusion, and we have learned from Macknik, Martinez-Conde, and their magical colleagues that misdirection plays a similarly important role in the performance of stage magic. We can even fit the type of misdirection at the heart of a performance of Who's Touching You? within the categories of directional tricks that stage magicians deploy; Who's Touching You? constitutes an example of *transposition*, in which an object magically changes position in space from position A to position B. In a 2008 article on attention and stage magic, a group of coauthors, including the scientists Macknik and Martinez-Conde *along with* magicians James Randi, Apollo Robbins, Teller, and Johnny Thompson, described four techniques of transposition, the first of which perfectly recapitulates the technique of Who's Touching You?: "The object seemed to be at A, but actually was already at B (for example, the magician fakes the transfer of a coin from the right to the left hand, then pretends to transfer the coin magically from left to right)" (Macknik et al. 2008, 874).[16] The reference here to professional legerdemain is important because unlike the magician's coin tricks and sleights of hand that require a level of practice and mastery characteristic of professional performance, agents in a performance of Who's Touching You? need no special training or trade secrets.

We might similarly think of Who's Touching You? as a less sophisticated version of the master pickpocket's techniques. Describing magician and professional pickpocket Apollo Robbins's performance at a 2007 Magic of Consciousness symposium held in Las Vegas, Macknik and Martinez-Conde capture the all-encompassing knowledge about attention that a master of the "kleptic arts" can deploy: "In the psychic sparring ring of attention management, Apollo is a tenth-degree black belt. By continually touching George [Apollo's volunteer selected from the audience] in various places—his shoulder, wrist, breast pocket, outer thigh—he jerks George's attention around the way a magnet draws a compass needle. While George is trying to keep track of it all, Apollo is delicately dipping his other hand into George's pockets, using his fast-driving voice to help keep George's attention riveted on Apollo's cognitive feints and jabs and away from the pockets being picked" (2010, 246). It is natural for us to protect that which is ours. Few things capture our immediate attention more powerfully than the threat of having our material possessions taken away. Ironically, these very natural tendencies to attend to things that seem important are a part of what makes us vulnerable. To some degree, agents in a folkloric performance of Who's Touching You? recognize, alongside the stage magician and pickpocket, that participants' attention and awareness can be manipulated for tricky, sometimes nefarious purposes.

If the professional magician's knowledge about misdirection is beneficial to the scientific study of attention and illusions—and we agree with Macknik and Martinez-Conde that it is—then we should ask whether the shared folk knowledge necessarily inherent to a performance of Who's Touching You? should also be beneficial. It is a difficult question to answer. For starters, our preceding discussion of Is Sammich Touching You?—a truncated and flawed performance of the illusion—points toward a set of decipherable, discrete variables that allow the patient to avoid being completely tricked, to avoid succumbing to the illusion. We listed these possible explanations for the flawed performance:

1. The agent's tickle on the patient's neck was too abnormal a type of touch to be actively perceived as social touch.
1a. The agent's tickle on the patient's neck was delivered in too abnormal of a body region to be actively perceived as social touch.
2. The patient had already noticed Sammich—an unusually large person—so his active perception of any social touch coming from that side was already influenced by attendant wariness.
3. The patient had previous knowledge of the agent's tendencies to engage in tricky behavior like a performance of Who's Touching You?, so his active perception of any social touch in the presence of the agent is always influenced by attendant wariness.

We also mentioned above that these are not the only possible explanations for the truncated performance, but each of the possible explanations constitutes a strong lesson for thinkers who concern themselves with the nature of active perception—especially active perception in social settings.

We end our discussion of active perception where it began, facing questions about the complicated merger of experimental and folkloristic data sets. Of the experimenter, we ask the following: Can any of the explanations above be isolated as testable variables? If so, are experimentalists willing to look past the triviality barrier and design such experiments? If not, why not? And for fieldworkers, we pose similar inquiries with a slightly different slant: What worth do we afford isolated treatments of the perceptual processes involved in prototypical performances of Who's Touching You? In the case that 1, 1a, 2, or 3 could be shown to affect the perception of social shoulder taps, would fieldworkers deem the findings less important (or, at least, less interesting) than the cultural backgrounds that make 1, 1a, 2, and/or 3 applicable to any particular performance?

Consider the folk illusions we discussed in this chapter; each plays with the sense of touch—surely a component of biological, universal innateness. Studied categorically, Crossed Fingers, Where Am I Touching You?, and Who's Touching You? provide folkloric commentary on the role that touch plays in youths' understandings of and theories about other aspects of embodiment like posture, proprioception, attention, and interpersonal touch. The illusions unveil traditionalized ideas about bodies, perception, causal relationships, and about the social environments in which they intersect.

Notes

1. References to Sherlock Holmes's methods are fairly common in nonliterary, academic writings. To our knowledge, Christof Koch and Tomaso Poggio were the first to use Holmes's exploits detailed in "Silver Blaze" as an analogical explanation for active perception—though they focus on a different episode within the story than we do. Referring to the dog that does not bark in the night, Koch and Poggio argue that predictive coding in visual perception also suppresses expected stimuli—like the silent dog in Conan Doyle's story who does not bark at the familiar, friendly horse trainer (1999, 9).

2. If you have plans to read "Silver Blaze" and *do not* desire to know just how the match features in Sherlock's solution, this is your spoiler alert! When he looks over the caked mud on the ground at the crime scene, Sherlock's running hypothesis is that Silver Blaze's trainer, John Straker, was not murdered. Instead, Sherlock theorizes that Straker had stolen the prize horse off into the dark night of the moorlands, so he could injure Silver Blaze's leg with a small knife prior to the next race. In fact, Sherlock believes that Straker's familiarity with the grounds is the reason why the stable's dog did not bark, which would have raised the alarm. Straker, having bet on Silver Blaze to lose, had much to gain from injuring the horse. In order to make the incision in the darkness of night, however, Straker would need a light source. The matchstick is a case-breaking clue. The horse, "frightened at the sudden glare, and with the strange instinct of animals feeling that some mischief was intended, lashed out, and the steel shoe had struck Straker full on the forehead" (Doyle 1976, 187).

3. Old homunculus problems have reemerged in contemporary philosophical descriptions of neuroscience; for recent arguments on the place of the anthropomorphic in neuroscience, see Bennett et al. (2009). On the trend of anthropomorphizing introversive—as opposed to extroversive—aspects of what it means to be human in writings on artificial intelligence, see Gregory Schrempp's "Descartes Descending or the Last Homunculus" in *Ancient Mythology of Modern Science* (2012, 151–90).

4. The most pertinent twentieth-century work on direct perception came from psychologist James J. Gibson (1904–79), who argued that certain invariant properties of objects in the environment (e.g., the actual size and shape of objects being visually processed) provide enough direct information to the viewer (despite the viewer's constantly changing point of view) that perceptual processing need not rely on cognitive functions to represent our physical environments. At least a portion of Gibson's work maintains supporters among twenty-first-century perception specialists—especially in regards to his theory of *affordances*. The notion of an affordance—which we revisit in chap. 5—is used frequently in sciences of perception and of mind in reference to the perceptual/conceptual possibilities that environmental stimuli afford an agent. For a short summary of sensist positions in Western psychological history, see Stanley Coren (2013, 102–6).

5. We were first reminded of this excellent reference to phantom limbs in Melville's masterpiece by our reading of Maclachlan's discussion of the history of "Illusory Body Experiences" (2004, 111).

6. For the original publication of the experiment and its findings, see Ramachandran, Rogers-Ramachandran, and Cobb's "Touching the Phantom Limb" (1995).

7. Gregory pulls especially from the work of statistician Thomas Bayes (1702–61) and from his eponymously named Bayes's theorem (sometimes called Bayesian probability). Bayesian probability—in contrast with objective theories of probability where probability is thought of as ratios of frequencies—espouses a subjective theory of probability in which prior probabilities are modified by present evidence just as the reliability of evidence is judged by prior probability.

8. We already introduced another folk illusion, Twisted Hands, during which children play with their confused body maps after assuming an irregular, contorted performance position (see chap. 2).

9. That is, when the subjects' fingers were displaced between 3 mm and 6 mm, Benedetti found sporadic reports of double-perception; however, when the finger pads were displaced between 9 mm and 12 mm, subjects consistently reported more double-perception experiences than single-perception experiences (1985, 87–88).

10. See Uncyclopedia, s.v. "The Oldest Trick in the Book," October 23, 2014, http://en.uncyclopedia.co/wiki/Oldest_trick_in_the_book.

11. Uncyclopedia, s.v. "The Oldest Trick in the Book."

12. Experiments on the spatial orientation of attention have shown a strong correlation between shortened response time to target stimuli when following a preceding exogenous stimulus that directs attention to the same side of the body. A typical visual study of symbolic orientation of attention involves subjects who are presented a computer screen with a box in the middle. Also on the computer screen are two boxes on either side of the middle box. Subjects are told that they will be presented a priming cue in the middle box prior to being presented a target cue in either the left- or right-side box. In this typical example, the priming cues presented in the middle box might take three forms: [X], [←], or [→]. Because attention orients toward a prompted side (i.e., the right or left side of the center of a forward-facing subject), subjects respond more quickly when a target cue presented in the right box follows a priming cue of [→] in the middle box. Similarly, oppositional priming and target cues lead to slower identification times. For a strong introduction to the study of spatial attention, see Chica et al.'s guide for experimenters, "The Spatial Orienting Paradigm" (2014).

For a particularly relevant study of attention and cross-modal interactions between vision and haptic sensations on the back, see Young, Tan, and Gray's 2003 study, "Validity of

Haptic Cues and Its Effect on Priming Visual Spatial Attention." In this experiment, touch sensations were delivered via a specially designed vibrating device draped over the back of an office chair that delivered a vibrating haptic sensation in one of four spaces. These haptic priming cues were followed by visual targets displayed on a computer screen. For valid haptic cues, a touch sensation was delivered on the subject's back in the same spatial position of the visually presented cue on the computer screen. So a touch sensation delivered to the lower left side of a subject's back would be followed by the visual presentation of a cue in the lower left corner of the computer screen. Invalid haptic cues involved any spatial mismatch between haptic and visual cues. Subjects were divided into two groups: one in which haptic cues were valid 80 percent of the time and one in which haptic cues were valid only 20 percent of the time. In the 80 percent group, subjects identified target cues 47 percent faster (as compared to visual identification and responses when no priming cues were administered) when priming cues were valid; response times after invalid priming cues were 7.5 percent slower.

13. *Change blindness* refers to an unnoticed change in a perceptual field. For Gregory, the questions of attention and change blindness are directly related to the pitfalls of active perception: "Change blindness is not surprising on the view that perceptions are *predictive hypotheses*. For hypotheses are useful for giving continuity through gaps of data. Of course, reliance on running hypotheses does sometimes let one down, but generally only occasionally topping up from fresh data is needed for continuous perception and continuous behavior, which is useful" (2009, 105–6).

14. The water crisis in Flint, Michigan, was a widely publicized failure of government and water-utility organizations to maintain safe water access for more than one hundred thousand people living in the city. For the original study of raised lead content in local Flint children's blood samples, see Hannah-Attisha et al. (2016).

15. Since the rise of performance-centered folkloristics, more attention has been given to truncated, unsuccessful instances of performance—especially when compared to the decentering of performance as imperfect or incomplete in Chomskyan linguistics. See Hymes (1975a, 11–13), and for commentary on the issue in the contexts of children's folklore, see Bauman's discussion of flawed knock-knock performances "as indices to the range of cognitive and communicative skills implicated in the performance of the genre" (1982, 178).

16. The authors also list the following methods for achieving the illusion of transposition:

- the object is still at A but seems to be at B (for example, the magician fakes a coin transfer from the left hand to the right and then, when revealing the coin by dropping it, uses sleight of hand to give the impression that it was dropped from the right hand)
- the object was secretly moved from A to B (for example, a coin in the left hand is secretly transferred to the right hand and then is revealed there)
- a duplicate object is used (for example, both hands hold identical coins that are revealed at different times to simulate a transfer) (Macknik et al. 2008, 874)

5

FOLK ILLUSIONS AND THE
WEIGHT OF THE WORLD

FOLKLORIC TRANSMISSION AND HUMAN PERCEPTION SHARE THE SIMILARITY of a dynamic, active core. This has been our argument for the preceding two chapters. Our goal in bringing the theoretical explanations of folklore and perception together, of course, is to illuminate the similarly active core of the genre folk illusions—in which covert properties of subjective experience and perception rise to the surface of social behavior, to the level of performance. As traditional performance, folk illusions signal simultaneously toward the communally constructed qualities of selfhood and the selfishly constructed qualities of community. In this chapter, we want to build on this foundational coalescence by examining folk illusions that play with an isolatable aspect of perceptual experience: namely, *weight*.

Weight surrounds us. It aligns our earliest notions of up and down, of heavy and light. It hangs on our bodies . . . and our minds. How heavy is the load we must carry? How light is our purse? And though it is a constant, weight perception often goes unnoticed. Japanese proverbial wisdom teaches that "one does not know one's own weight"—a metaphorical suggestion that we remain unaware of the breadth of our positive *and* negative influences.[1] Among the Xhosa and Zulu people of southern Africa, tradition maintains that "a mother does not feel her baby's weight"—a sweet premise that forces real-world retrospection: If weight is not felt, does it even exist?[2] Then again, whether or not we find ourselves in the midst of existential crisis, everyone—we say in English—must manage to get around with "the weight of the world on our shoulders." Formulaic phrases of this sort point to the fact that human experience of weight is not relegated to bodily interactions; we internalize, construe, and project our notions of weight into our social worlds of communication and tradition where paradox abounds. In those folkloric spaces, the everywhereness of weight splits into familiar-yet-strange patterns of differentiation—that is, not all acts of floating, falling, lifting, sinking, or levitating are one and the same.

Weight is an appropriate phenomenon to consider within the genre of folk illusions if for no other reason than the fact that weight plays a central role in several forms we have gathered and cataloged. If you have not yet perused the catalog of folk illusions in the appendix at the end of this book, this is another good point at which to do so. Especially relevant to our analysis of weight are the thirteen forms that fall under sections F, "Weight Illusions," and G, "Falling and Flying Illusions," as they deal directly with the perceived weight of an object or of the performers' own bodies.[3] We are motivated in this chapter, however, by more than weight's prevalence in our catalog. Interestingly, the perception of weight has played a significant role in the study of perceptual illusions. Inasmuch as an object's weight can be verified via simple (and complex) instruments of technology, human interactions with weight lend themselves to scientific reports on both the objective and the subjective. A twenty-pound weight can be accurately measured as such on countless scales, but this does not mean that people's experiences of lifting that weight are the same. And while it goes without saying that lifters' strength levels or their states of relative fatigue affect the experience of hoisting up a twenty-pound weight, in this chapter we want to examine the ways that social frames and performances realized as folk illusions affect the perception of weight.[4] Do folk illusions affect youth's idea of weight, and vice versa?

For comparative psychologist Daniel Povinelli, people's interactions with weight demonstrate humans' inherent, unmatched propensity for theory-driven reasoning about causal relationships out in the world. The following passage comes from the book-length study by Povinelli and his colleagues of the *absence* of a theory of weight in the minds of humans' closest biological relative, the chimpanzee:

> In every domain of reasoning—from time and space, to mental states and physical illness—humans deploy an exceedingly diverse range of intuitive "theories." To be sure, children from diverse cultures always seem to arrive at a few, common folk theories as they hone their developing brains against roughly similar interactions with people and objects. But as their experiences diverge, widely disparate cultural narratives shape and reshape those theories, transforming them and reforming new ones. There's a great deal of debate in cognitive science about whether our folk ideas about the world and each other are largely "innate" or largely "learned" or largely something in between.

> But one thing's for certain: as Richard Nisbett and Timothy Wilson showed
> many years ago, our species is primed to generate explanations for events—
> regardless of whether we have the requisite background information to do so.
> The result is an impressive panoply of folk notions that we use to explain, pre-
> dict, and just plain talk about everything from why the sky is blue to why we
> catch a cold when we stand out in the rain. (2012, vi)[5]

While it will remain important to recognize the differing inflection—as
Greg Schrempp describes it—between the scientist's use of *folk* and the
folkloristic use of the term (see our discussion of *folk*, chap. 1), Povinelli's
nod toward an impressive and enabling "panoply of folk notions" rings
true across the science/social-science divide.[6] And as far as "disparate cultural narratives" are concerned, folklore—as we have argued throughout—
magnifies important intersections of the "innate" and the "learned," regardless of precisely where we draw the nature-nurture line. In folk illusions, we find that people's (folk) theories of weight are wrapped up in the
ways that they play with illusory weight(s), and that those play forms demonstrate weight's role in the creation of their worldviews.

Heavy Is the Head, or Mind-Body Representations of Weight

From seemingly helpless newborns, children are upright and teetering
about within the first two years of life. Developmental scientists Eleanor J.
Gibson and Anne D. Pick, who take an ecological approach to perception,[7]
remind us that many aspects of physiological and cognitive development—
especially postural control—precede the child's first precarious steps:

> A shift in proportions as the child grows less top heavy and stronger in the legs
> is among the factors that make walking possible. The gradual achievement of
> control of posture underlies all forms of perception-action development.
>   Indeed, postural development is a leading factor in all behavioral
> development.
>   Individual actions, such as an infant lifting a hand to the mouth, al-
> ways occur against a postural background and, in fact, can occur only to the
> extent that the infant is supported and posturally stable. Differentiation of
> actions occurs as postural control develops, beginning with control of head
> and shoulders and proceeding downward, depending on growth of limbs and
> many other factors. As new actions become possible, infants' potential for ex-
> ploring their surroundings and themselves grows. (2000, 23)

Humans' first actions that deal with weight, then, are the actions that deal
with the weight of the self. Babies can usually support their own head within

two months and can sit up on their own within five to seven months. By the end of the first year, they usually achieve some form of locomotion (crawling, scooting, etc.), and as anyone who has *tried* to keep a crawling baby out of dangerous spaces knows, the ability to crawl immediately affords opportunities to play with the weight of a seemingly endless amount of now-obtainable objects like toys, scissors, important papers, and car keys!

Mounting opportunities for children to interact with the weight of the world only increase with development. From getting dressed in the morning to putting food in our mouths, the responsibilities that accompany even our earliest versions of self-reliance require that we interact with weight in countless ways. Psychologists and cognitive scientists generally split our dealings with weight into two categories: sensorimotor and cognitive. Sensorimotor processes are thought to include all of the perceptual and motor actions that deal with the weight of objects (including our own bodies) without the need for conscious attention. To demonstrate just how much information we deal with outside of consciousness, neuroscientist Frank Wilson uses the example of a worker carrying a bag of cement on his shoulder while traversing up a flight of stairs in order to finally dump the bag of cement into a mixer:

> Each time he takes a step he is actually throwing himself off balance by shifting support of the load from one leg to the other. He does the same if he leans slightly to one side or the other, or shifts the bag from one shoulder to the other, or repositions the bag vertically or laterally anywhere along his shoulder or arm. If he steps on a wet or dusty surface his foot may slide, and he must instantaneously restabilize the load or risk falling. At the same time, he must be aware of internal sensations arising from his musculoskeletal system: if he exceeds the load- and tension-bearing capacities of his own muscles, bones, joints, and ligaments he may seriously injure himself.
>
> The problem is so complicated for the neuromuscular system that the perceptual and control side of the job is handled almost entirely outside conscious awareness. There is far too much information about joint angles and weight distribution to be gathered and analyzed, and there are far too many inertial and force equations for the brain to solve rapidly and simultaneously. (1998, 64–65)

In Wilson's description, the "perceptual and control side of the job" that is "handled almost entirely outside conscious awareness" involves sensorimotor processes. Put in these terms, these unconscious interactions with weight seem daunting, if not amazing, mechanical and physiological feats.

We need not completely unpackage the representational complexities of the sensorimotor perception of weight, however, to recognize that people also often deal with weight consciously and cognitively. We are especially attentive to an object, for example, when its weight surprises us. Think of the last time you observed an adult family member, an aunt perhaps, pick up her niece only to stagger under the weight that growing children add so quickly. The aunt's benevolent struggle signals the surreality of just how fast our little ones grow up, but it also exemplifies the fact that the inner Sherlock keeps track of our interactions with weight, over time and across specific situations. When our hypothetical aunt's expectation for just how much her little niece will weigh proves to be incorrect, she easily translates the phenomenological experience of weight into verbal representation: "Wow, you are getting heavy!"

Remember now Povinelli's work on specific domains of causal reasoning, such as weight. Sensorimotor and mental representations of weight are reinterpreted in our *folk theories* concerning "the common causal effects of 'weight' across disparate perceptual contexts"—a characteristic of human minds that he names *function weight,* or *f*(weight). Children develop something very close to a full theory of *f*(weight) between the ages of three and five, so you can easily call *f*(weight) to mind.[8] As you read these words, you are concentrating—we suppose—on the issues of weight that we have just introduced, but you are also viewing our text. Perhaps it is printed in a book or displayed on an electronic tablet? Maybe you are reading on your computer. Is this computer a laptop or a desktop? Whichever reading device you prefer, we can be certain that you have theories concerning its weight.[9] You have expectations, for example, as to what kinds of sound (tonal quality, loudness, etc.) each one of the reading technologies would make if dropped onto the floor. The laptop's plastic crunch would be quite different from the book's holistic thud. A tablet, you theorize, wouldn't be nearly as loud as your desktop. The same theory (or group of theories) about weight allows you to predict which of the reading technologies would be better suited for smashing through a stuck window in, say, the case of a fire. Likewise, if each of these reading technologies were flying at your face *and* if you could dodge all but one, you would not choose the destructive desktop computer, for you understand the weight of a flying object is directly related to the amount of pain it can cause. Hence, the delightful (if childish) logic of Acme safes falling from the sky!

How the Human Body and Brain Represent "Weight"

Adapted from Daniel Povinelli (2012, 7)

(1) Antigravity muscles	The mass of many muscles in the body quickly atrophy in response to weightlessness.
(2) Effort required to lift object	The sensorimotor system tracks the force used to lift a given object.
(3) Effort experienced in lifting object	Humans (and other animals) have phenomenological experiences of effort in lifting and moving objects.
(4) Feed-forward model of effort required	The sensorimotor system represents the predicted-to-lift-object effort needed to lift a given object.
(5) Causal representation of a relation involving "weight"	The perceptual system represents the causal connections between object and outcome (organisms may use size, weight, shape, effort sensation, or any other cluster of these to learn to predict an "effect" from a cause).
(6) Effects of object weight across causal contexts, or f(weight)	The human mind represents "weight" as a mechanism common across perceptually distinct categories.
(7) Linguistic representation	Humans generate linguistic (symbolic) labels for the causal mechanism of weight ("weight," "heavy," "light," etc.).
(8) Metaphorical expression	Humans generate linguistic (symbolic) descriptions that connect both the phenomenological aspects of weight and f(weight) to nonliteral contexts.

Cartoons present only one small portion of the relationship between children's culture and weight. Recalling Wilson's hypothetical worker carrying a bag of concrete, we might think of children's play like diving into swimming pools and jumping on trampolines, skipping, bouncing balls, jumping rope, flying kites, and holding helium-filled balloons as activities that play with related phenomenological and mental processes—and with f(weight). How much youthful play breaks through the weight of kids' otherwise tediously grounded ecologies?

The Not-So-Heavy Object and the Mind-Body Distribution of Weight Illusions

In the complex of cognitive expectations and mental theories about f(weight) coupled with sensorimotor manipulation of actual objects and body weight, it is clear that weight perception is distributed across many mind-body processes. Take, for example, weight and memory. Turning experience into expectations, we rarely—if ever—pick up an object without our inner Sherlock having some preconceived notion of the object's weight, and weight illusions show us that cooperation across the involved embodied systems is not always seamless. Surprisingly, we find a scientific reference to folkloric awareness of this obvious fact in a 2001 article about experience and object manipulation by neuroscientists Randall Flanagan, Sara King, Daniel Wolpert, and Roland Johansson: "Most of us will recall having fallen victim to a mischievous older sibling, cousin, or dubious friend who passed us an empty box while pretending it was very heavy. When we took the box, and the bait, our arms flailed upwards. This trick demonstrates that when we interact with objects, we anticipate the forces required to accomplish the task at hand. Although it may occasionally result in large movement errors, predictive or anticipatory control is essential for skilled object manipulation, whether wielding a tennis racquet or simply picking up a cup" (2001, 89).

The scientists, here, echo our central claim that people out in the world are well attuned to many illusory processes; weight illusions undoubtedly fall in this category.[10] The important feature of the illusion in this tricky tradition—a folk illusion we call the Not-So-Heavy Object—is often referred to as the positional overshoot effect or empty-suitcase phenomenon, which coincides with the application of too much grip-force and lift-force.[11] Flanagan and his colleagues' remembrance of "arms flailing upwards" captures the most important quality of the positional overshoot in the social contexts of folk illusions; specifically, it is an observable phenomenon. Pranksters get to watch their targets succumb to the illusion.

The idea behind a performance of the Not-So-Heavy Object—an idea closely related to the idea behind experimentally derived weight illusions—is that an object can be made to feel heavier (or lighter) than it actually is. We will go to some length in this chapter to introduce the scientists' awareness of this idea, but what of popular cultural awareness? One quick way to gauge an idea's presence in popular culture is to check for commodification of the idea. If you search, for example, Amazon.com's website for "fake brick," you will find several products that capture the kind of weight predictions described above by Flanagan and his colleagues. When we entered "fake brick" into Amazon's search engine in June 2016, one of the first options we found was NPW's gift item, Bricking It!, "fake foam brick with impact activated sound."

Fig. 5.1. NPW's novelty item, Bricking It!, combines onomatopoeic sounds with stunning visual accuracy. (Printed with permission of NPW.)

With the red look of a clay brick, Bricking It! presents itself to the casual onlooker as an actual brick. Here is NPW's bullet-point description:

- Novelty faux brick emits the sound of breaking glass on impact
- Approximately 7.75 × 2 × 3.75 inches
- Foam brick can safely be hurled across a room
- Buy several to throw around at work meetings or watching the game on TV
- Entertain friends with hilarious NPW gifts and gadgets that will crack them up

Key elements of folk illusions highlight NPW's list. The first and most obvious being the brick's intended use, which is, of course, to create a playful social scenario by tricking some participant's perception.

It is telling that NPW's list never refers directly to perception or weight, because the fact that the foam brick tricks people's intuitive understanding of f(weight) is obvious. The multisensory illusion arises from prior

Fig. 5.2. Mr. Martin's great-granddaughter, ten-year-old Ava Loewer, lifts the "sledgehammer" with ease.

experience with heavy bricks that can easily crash through a pane of glass. We have gathered remembrances recounting avid sports fans who kept low-tech versions of the fake brick by their favorite chair to throw at the television in reaction to a bad play and of sly teachers who kept such a fake brick to throw at unruly, unresponsive students.[12]

Brick It! differs from the Not-So-Heavy Object, however, in important ways. First, the manufacturing processes associated with Brick It! and other products like it allow for deliberate, complex technologies to create highly realistic fakes.[13] Second, the respective performances (at least, the range of performances suggested by NPW) do not seem to necessarily involve a naive actor lifting the lighter-than-expected object.

In 2016, we came across a traditionalized, material-culture example of the illusion that combines Brick It!'s fidelity of replication with the-Not-So-Heavy Object's intended positional overshoot. The apparent sledgehammer pictured in figure 5.2 was made by ninety-two-year-old Samuel Martin some thirty years ago. A Pennsylvanian German woodworker, he is especially fond of asking unsuspecting friends and family to go and pick up the hammer for him, so that everyone in the room can laugh at the positional overshoot! Mr. Martin, a sharp and playful storyteller, recounts his construction of the hammer:

> It's solid wood, but it came from a dead limb, laying around for I don't know how many years. Then, one day I pick this thing up. It has no weight to it! And then I had a friend of mine at a sawmill, he was sort of a tricky guy. He had a hammer like that. It didn't weigh much. I don't know what wood he had.
> And I decided to make a hammer out of it. It just don't have no weight; it's like a honeycomb on the inside. And yet when you sand it, it gets nice and smooth. So, one day I thought, hey, I'm going to have some fun. I made a hammer. And then I bought a hammer handle, and I took a long bit, and I drilled it about a foot and a half at least to get it lighter yet. There's just no weight to it. Maybe a pound, but not more. . . . No, I don't believe it'd be even that.

Because the hammer can always be found resting against the wall just inside the doorway, Mr. Martin told us, anyone who wants to pick it up can succumb to the trick. But as we mentioned, he sometimes chooses to play the role of director and prompts the children to pick it up: "And I tell you what, I'll have it sitting by the door there. I'll ask the gal to pick it up, 'Hey, girl, pick up that heavy hammer' . . . Whoop, nothing to it!" Prompting the child, Mr. Martin merges the intersubjective awareness of f(weight) with strong social performance. Sledgehammers are used to bash and crush strong solid materials, so they need to be heavy. Even the child knows this.

At other times, Mr. Martin's devices reveal complex insights into the nature of adult perception, such as the fact that culturally governed environments prepare us to interact with familiar objects: "I have fun with it. At the hardware store one time . . . I went to the hardware store with my hammer. And I had a price tag on it, and I walked up to the counter. The clerk says, 'Boy, you're really going to do some working!' And he grabbed a hold of it to pick it up. . . . I really fooled him!" In his hardware-store variant, Mr. Martin brandishes a strong awareness of the statistically formulated hypotheses that govern perception. Ask yourself, what is the probability that something looking just like a sledgehammer is heavy? Now ask yourself, what is the probability that something that looks just like a sledgehammer in a hardware store is heavy? Now, pretend to be the poor clerk!

Which Is Heavier?

An implicit cause of the positional overshoot effect is the comparative stance from which our inner Sherlock creates expectations about any given object's weight. We do not experience weight—be it the weight of our body, of a can of soda, of the spears with which we hunt, of the bicycle we ride, or even of a feather from our pillow—in isolation. Mr. Martin's sledgehammer is only illusory in comparison to all of the nonillusory sledgehammers that precede it. Illusions, it seems, need reality.

Decipherable quirks surrounding the inherently comparative nature of weight perception have driven the study of weight illusions for more than a century, and it is worth mentioning that scientists have discovered several scenarios in which objects of the same weight are judged to have differing weights. Most famously, size-weight illusions—which we first introduced in chapter 2's discussion of Light as a Feather, Stiff as a Board—have been studied since the end of the nineteenth century when French physiologist Augustin Charpentier (1891) found that subjects who lift two objects of differing size but of identical weight consistently judge the smaller object to weigh more than the larger one. Cognitive scientists understand the size-weight illusion to be "cognitively impenetrable": even if people are allowed to examine the objects and weigh them on a scale for themselves, they cannot escape the illusion. It is so powerful that apparently it comes with its very own sampling of in-group folklore in which scientists report "having to frequently reweigh their test items because they themselves experience the illusion as they are preparing to test their subjects!" (Povinelli 2012, 35).

You do not need highly specialized laboratory implements to experiment with the size-weight illusion. Richard Gregory describes a homespun version that utilizes common kitchen supplies: "Get two tin cans (or cans), one large, the other small. Fill them partially, for example with sugar, until they are the same scale weight. So we have large and small objects of the same weight. Do they feel the same weight? Try lifting them, one in each hand, or one after the other with the same hand" (1998, 221). In Gregory's version, the smaller can feels heavier than the larger can. He offers his now-familiar hypothesis-driven perception explanation for the illusion: "Larger objects are usually heavier than smaller objects. Although this predictive perceptual hypothesis is usually appropriate, here it is not—and we have a very suggestive shared cognitive illusion" (222).

To set up our favorite version of an experimentalesque size-weight illusion, you will need to acquire three boxes of playing cards. Remove all of the cards from the boxes. Then, fill one of the three boxes with two rolls of pennies (or anything else that will fit and that weighs in at just over nine ounces [i.e., three hundred grams]—the approximate weight of three decks of cards). Now, set the penny-filled box on top of the other empty boxes. Lift all three boxes with one hand, gripping the sides of the stacked boxes. Finally, lift only the top box (the one containing the pennies) with the same hand, still gripping the sides of the box.

How can the top box weigh more than the entire stack of three boxes? Try the lifts several times. As we mentioned, the size-weight illusion is notorious among scientists for being cognitively impenetrable.

Earliest explanations of the size-weight illusion focused on a mismatch between cognitive expectations and actual felt weight. According to these theories, you (or your inner Sherlock) would expect the stack of three playing-card boxes to weigh significantly more than the single top box, and the expectation for a lighter single box tricks you into applying too little grip- and lift-force, making the single box seem heavier than it really is. It is easy to see why such theories are tempting. The positional overshoot clearly demonstrates that some form of planning for grip- and lift-force occurs; no doubt, we do expect certain objects to weigh certain amounts.

But the situation is more complicated than this. Findings from recent experiments suggest that discrepancies between top-down (cognitive) and bottom-up (sensorimotor) processes cannot account for the entire phenomenon of size-weight illusions. Via a clever setup that allowed specially designed instruments to measure grip-force, the aforementioned Randall

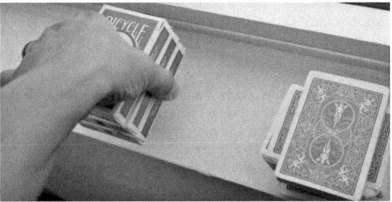

Fig. 5.3. The top box feels surprisingly heavier than the total weight of the stack of boxes.

Flanagan and Michael Beltzner showed in 2000 that lifters who alternate lifting smaller and then larger objects in succession (in approximately five to ten lifting trials) scale their grip-force to the *actual* weight of the lifted object, which means that participants eventually applied the same amount of grip-force for each object—large and small. Perplexingly—and this is perplexing—subjects continued to report the experience of a size-weight illusion even after their grip-force was correctly scaled. Flanagan and Beltzner react to these findings by offering an alternative to top-down/bottom-up mismatch theories: "Our finding that the size-weight illusion was not associated with errors in sensorimotor prediction argues for a purely cognitive/perceptual account" (2000, 740). It is a powerful argument. If the sensorimotor systems of the body "figure out" the correct and necessary forces required to lift an object, there can be no mismatch between bodily expectations and the actual lift. Yet the illusion remains, a fact that begs for a more nuanced explanation of the experience.

Flanagan and Beltzner's study suggests that future explanations for size-weight illusions will need to expand the role of cognition and of the mind in the perception of weight. We have come to think of Flanagan and Beltzner's study as experimental commentary on the worth of ethnographic paradigms in the study of illusory perceptions. Without the subjects' reports, the subjective experience of the illusion cannot be accessed by the scientists. Remember, the instrumentally derived force-grip data indicate that no illusion has been experienced. For us, the inclusion of discursive processes in experimental investigation parallels the folklorists' in that the scientist *and* the fieldworker take these verbal reports as genuine representations of the phenomenological experience.

Enculturated Expectations and the Golf Ball Illusion

While the size-weight illusion remains a central topic in the experimental investigation of weight illusions, it is important to note here that scientists have also noted and studied weight illusions arising from object properties separate from size-weight ratios. These include, for example, the (surface) material-weight illusion, the brightness-weight illusion, and the temperature-weight illusion.[14] Even culture has come to play an important role in scientists' understandings of weight illusions. In a particularly relevant study, psychologists Robert Ellis and Susan Lederman isolated a laboratory situation in which golfers' knowledge about golf ball varieties could be engaged in such a way as to produce size-weight illusions. "In the game of golf, there are two main types of golf balls. Real golf balls conform to strict specifications. They weigh 45 g. Practice golf balls are very similar to real golf balls but are hollow and weigh only 7 g. Because of their low weight, practice balls do not go very far no matter how hard they are struck. This characteristic makes them ideal for backyard practice. In other respects, particularly their outer covering (a hard, dimpled, white plastic), they are nearly identical to real golf balls" (1998, 196).

Because the plastic and dimpled material-surface characteristics of practice and real golf balls are the same, the only visually observable differences are superficial color or letter markings associated with particular brands. These surface characteristics were maintained, but the weight of the golf balls was manipulated so that both the "practice" balls and "real" balls weighed the same. Then, golfers and nongolfers were asked to lift and estimate the weight of the two sets of golf balls—without knowledge of the scientists' behind-the-scenes manipulations. Golfers experienced a size-weight illusion, reporting that the "practice" balls weighed more than the "real" balls. Nongolfers, on the other hand, did not experience an illusion and reported that the two sets of "real" and "practice" balls weighed the same.

In the Golf Ball illusion, emic knowledge directly affects perception—a specific and powerful example of how weight perception is both biologically *and* culturally governed. No matter where we draw the lines between sensorimotor and cognitive processes, golfers' highly enculturated experience with the weight of their sports equipment rises to the surface as the key determiner of whether or not the illusion occurs.[15]

Combine Our Strengths Illusion

The Golf Ball illusion relies on a particular cultural frame for the perception of weight, but no version of the experiment, as far as we know, has made its way into children's traditional culture. Like the professional magicians we discussed in the previous chapter, the amount of preparation and training required of the scientists who created the Golf Ball illusion surpasses the amount of preparation typical of folk illusions' performers. We do, however, know that children and youth play with overt comparisons of object weight in other ways. The best example is a variant of levitation play, which we call Combine Our Strengths (F3a in the appendix). We first learned of Combine Our Strengths on the playgrounds of St. Cecilia Middle School in 2011.[16] In this performance, four seventh-grade boys gathered around their science teacher, who was seated in a chair. The directors (i.e., the boys)—using only two fingers—lifted the actor (i.e., their teacher) from the chair and marked the height of the first lift. Then, the directors alternately stacked their hands on top of the actor's head and lifted the actor a second time. Undeniably, the second lift was much higher than the first!

Just as the teacher was safely lowered back into her chair, one child proclaimed, "It's a fact, it's scientifically proven," despite the obvious fact that he did not relay the exact details of the scientific principles. Likewise, none of the performers mentioned the scientific principles of weight distribution in relation to the illusion. When the boys were asked why the second lift goes higher, theories and pseudoscientific explanations came flying in from every direction.

"Because, every time you do this!" [mimes stacking hands]

"You didn't expect it the first time, but you expect it the second time!"

"We don't know. It just works, we figured it out."

"Joining hands [mimes stacking hands], gives each other strengths!"

"Like if you put a doughnut [a weight] on a [baseball] bat, it gives you more, it makes it lighter pretty much."

The boys' justifications gesture toward scientifically grounded explanations like weight distribution and increased muscular performance following resistance exercises. Indeed, these nearly scientific explanations for the higher second lift seem to be a key traditional element of the variant. In 2014, a twenty-seven-year-old male who grew up in Venezuela explained to us that the hand-stacking elements actually "cut the center of gravity." As off-base as some of these theories might seem (how can a center of gravity be cut?), it is important *not* to think of the explanations of the seventh graders as incorrect theories (scientific or otherwise), because relegating the youth's theories to the world of the confirmable would miss key insights.

We learn strategic elements of youthful theory-creation, for example, by comparing Combine Our Strengths lifts to those of more ritualesque traditions like Light as a Feather, which we introduced in chapter 2. Remember that the necessary contexts for the creation of Light as a Feather's lift are similarly distributed because the participation of the many lifters is necessary to achieve the illusion's success. For levitation to occur, the group must assume their performance positions so that they properly distribute the actor's weight. Do they assume this position without recognizing the physical principles underlying weight distribution? Consciously, the

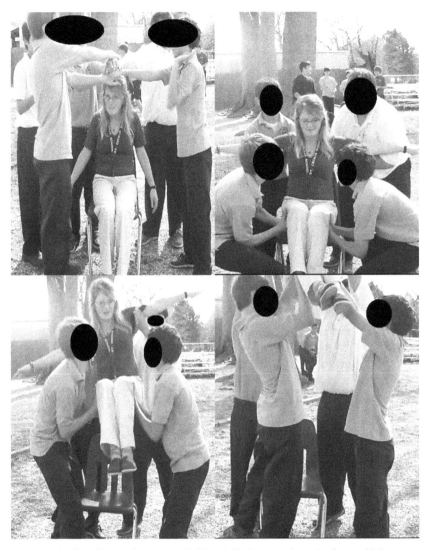

Fig. 5.4. After "combining their strengths" (*top left*), the seventh graders' second lift was significantly higher.

children are certainly focused on the macabre, spiritual aspects of the performance and the social flow of the moment. These supernatural, séance-like elements constitute the cultural frame that the illusion relies upon. Among all this—that is, the material and physiological contexts of the lift *and* the top-down, cultural frame of supernatural presence—some youths report that the actor supernaturally levitates. Like the scientists who believe the golfers' reports that the "practice" golf balls are heavier than the "real" golf balls, we take the children's reports of levitation as genuine descriptions of the phenomenological experience.

Juxtaposing Light as a Feather and Combine Our Strengths, we find that participant roles remain relatively stable while performance positions and the mechanics of the lifts reveal some variation. In Light as a Feather, the patient is lying down while the patient in Combine Our Strengths is seated in a chair. The patient is lifted only once in a performance of Light as a Feather. In Combine Our Strengths, the phenomenally askew lift (the levitation illusion) is actually a second lift that goes higher, showing that the burden on the lifting participants had been lessened. These differences are ancillary, however, to the *raison d'être* of the performance: lifting a playmate higher than would be possible without the assistance of an unobservable, uncanny force. The important explanatory differences appear in the forms' respective verbal components and performance contexts: a séance-like context with a call to unseen spiritual powers *or* an experiment-like context that harnesses unseen, mysterious, scientifically founded powers.

Even in their variation, however, these verbal components and performance contexts remain culturally restricted since they must promote communal belief in the actuality of the unseen, mysterious forces that allow levitation to occur. Thus far, we have not learned of a levitation play variant that credits the lifting of the actor to invisible aliens, though we can imagine such a variant. One reason why we have not come across such a variant is the fact that we have not fieldworked a group of children for whom such a theme would promote belief in the actuality of levitation. Unlike invisible aliens, the belief in complex, unseen scientific phenomena and the belief in powerful, unseen spiritual forces are both living folk beliefs among the students at St. Cecilia.[17] During a successful performance of complex variants of levitation play like Light as a Feather and Combine Our Strengths, children infer not only that their playmates are experiencing the felt, proprioceptive aspects of the illusion but also that their playmates believe that their experience is the result of the unseen forces suggested by the accompanying belief structures.

Where the Golf Ball illusion relies on an individual's previous experience with culturally modulated objects (practice and game golf balls), the Combine Our Strengths illusion arises from group experience with all kinds of weights . . . and with each other. The discursive aspects of their communal experiences reify preexisting beliefs in the unseen, mysterious powers of science and the supernatural. The physiological experience of the size-weight illusion and the socially manifest beliefs prop each other up.

Internal Perceptions of Body Weight and the Illusion-Weight Illusion

To this point, we have argued that folk illusions like the Not-So-Heavy Object, Light as a Feather, and Combine Our Strengths play with youths' theories about weight and with their inner Sherlocks' predictions about weight as much as they play with the physiological activities involved in lifting objects out in the world like hammers, bricks, and other people. Now, we move to an even more clearly introspective category of weight perception in folk illusions that defamiliarize the weight of the self—the subjectively perceived weight of one's own body. We have already introduced one such example.

Floating Arms, introduced in chapter 2, defies even youthful logic about how the weight of our arms should or should not be perceived during normal material, physiological contexts. Arms should rest at a child's side unless the child or some other agent acts upon those arms. But Floating Arms defies this logic; the arms seemingly lift on their own. As a folk illusion and a subject of scientific investigation, fascination with this Kohnstamm effect has seemingly always been directly related to questions about why the arms lift. As early as 1921, two scientists reported that subjects always have the feeling that a foreign force beyond the subjects' control is lifting their arms (Schwartz and Meyer 1921, 490). Youthful performers are left questioning their arms: Who lifts them? How are they lifted?

Scientists, it turns out, ask their subjects much the same questions. In an excellent 2016 study of voluntary and involuntary motor control, neuroscientists Jack De Havas, Arko Ghosh, Hiroaki Gomi, and Patrick Haggard induced basically the same Kohnstamm effect that youths play with in a performance of Floating Arms in one arm of fourteen subjects. More germane to our discussion of verbal and cultural frames for weight perception is the set of forty-six questions that gauged the participants' subjective experiences. We reproduce the entire list as well as the relevant numerical data in a box here. In response to the questionnaire, "[p]articipants rated each statement from −3 (strongly disagree) to 3 (strongly agree) on a 7-point Likert scale" (159).

The scientists interpret the results of this questionnaire as confirmations of "previous reports" on the experience. Specifically, answers to (4),

Kohnstamm Survey Question List

From Jack De Havas et al. (2016, 161)

(1)   The movement seemed to begin automatically
(2)   My arm seemed lighter than normal
(3)   I found the experience of my arm moving interesting
(4)   I had to will my arm to begin the movement
(5)   It seemed like gravity was not acting on my arm
(6)   I found the experience of my arm moving boring
(7)   The rest of my body felt normal during the movement
(8)   It seemed the movement was involuntary
(9)   It seemed like my arm was being buoyed up by water
(10) I found the experience of my arm moving pleasant
(11) I had the sensation of pins and needles in my arm
(12) It seemed like a cushion of air was lifting my arm
(13) It seemed like my arm was being pulled upwards by a rope
(14) My arm seemed heavier than normal
(15) I felt a greater sense of freedom during this arm movement than normal movements
(16) I found the experience of my arm moving frightening
(17) It seemed like my arm was being lifted by a helium balloon
(18) The experience seemed dreamlike
(19) I experienced a sense of relief when my arm started to move
(20) The movement seemed smoother than normal movements
(21) It seemed like my arm was full of helium
(22) I felt like I could control the speed of my arm
(23) It seemed like I was in control of the moving arm
(24) I knew where my arm was during the movement
(25) As my arm began to move I had the sensation that it would not stop
(26) It seemed like my arm was out of my control
(27) I found the experience of my arm moving strange
(28) It seemed like the moving arm did not belong to me
(29) It seemed the experience of my arm was less vivid than normal
(30) I had the sensation that my arm was numb
(31) It seemed like I couldn't really tell where my arm was in space
(32) I had to keep telling my arm to stay still
(33) It seemed like my arm was pulled upwards and I was pulling against that force
(34) When I stopped my arm I felt like upward drive was put on hold
(35) It was a relief when my arm stopped moving
(36) When my arm was stationary I had an urge to allow it to move again
(37) I only had to tell my arm to stop once and then it did not move

(38) When stationary, it seemed like my arm was resting on a cushion of air
(39) It was difficult to maintain my arm in a stationary position
(40) When I stopped my arm I felt like upward drive ended
(41) When stationary, it seemed like my arm was resting on water
(42) I found it easy to make my arm stop moving
(43) When stationary, it seemed like my arm was resting on a solid object
(44) It was easy to stop one hand without affecting the other
(45) When I stopped one hand the other hand also briefly stopped
(46) This task was easier than doing the same task with voluntary movement

(8), and (23) show that the Kohnstamm aftereffect was experienced as involuntary, while question (1) supports the idea that the effects are automatic. Responses to (9), (12), (13), and (17) correlate with the subjects' feeling of a lack of agency. And (2), (5), (14), and (21) represent the feeling of lightness in the arms commonly associated with the activity. They add that questions (32), (36), and (37) offer support for a somewhat lesser-known quality of the experience that voluntary inhibition of the Kohnstamm aftereffect had to be continuously and consciously engaged and was accompanied by an urge to allow the arms to float up.

We notice the direct parallels between the scientists' wording of their survey questions and bona fide folkloric descriptions of the experience we have gathered in our surveys and fieldwork. We have collected, for example, remembrances of the idea of automaticity in question (1), "The movement seemed to begin automatically," from three respondents. A folklore student from Valparaiso, Indiana, performed Floating Arms in gymnastics class in third grade: "The Floating Arms illusion, where someone pushed down on your arms for an amount of time then when they let go *your arms automatically came up*" (emphasis ours). Another folklore student played Floating Arms at his elementary school in Cherry Hill, New Jersey: "Arms rising to the sky, put your arms in your pockets and try to move your arms away from body creating tension on your pants, continue for 30 seconds, when you remove your hands, *your arms will automatically raise to the sky*" (emphasis ours). A third student also played Floating Arms in her elementary-school years, but she remembers playing at a friend's house: "You push your arms against a door frame for 30 seconds and then *your arms will automatically raise*" (emphasis ours).

These references to the automatic nature of the experience during folk-loric performance demonstrate, on the one hand, the stability of embodied processes across performative contexts—both in and out of the lab. But, on the other hand, they open the doorways to variation. The verbal descriptions above correlate with three varying physiological processes for creating the folk versions of the form: a director holding down the actor's arms, the actor placing his hands in his pockets, and an actor pushing her arms on the outside of the door frame.

Question (17), "It seemed like my arm was being lifted by a helium balloon," reminds us of Ramachandran and Ramachandran's reference to their son, who taught his parents how to make their arms float, "as if being lifted by helium balloons" (mentioned in chap. 2). In the spring of 2015, a nineteen-year-old female student from St. Leon, Indiana, described the activity like this: "Hold [your friend's] arms down, and let them go so they *feel weightless*" (emphasis ours)—a clear folkloric correlate to questions (2) and (5). Another student who performed the folk illusion as a twelve-year-old in Cincinnati, Ohio, provided a description that nearly exactly matches the description of question (5), "It seemed like gravity was not acting on my arm": "I played at my Catholic school on the playground. I forgot the name, but we said *the gravity was taken away from our arms*" (emphasis ours).

In the late summer of 2016, we contacted De Havas about this interesting study and questionnaire. As has been the case throughout our foray into the workings of experimental scientists, we were met with interest and kindness. He was immediately receptive and helpful. De Havas agrees that culture clearly plays a role in the descriptions of the causes of the Kohnstamm involuntary movement. As a matter of fact, the similarities between descriptions featured in the scientists' questionnaire and the descriptions we have received in our surveys and fieldwork might occur precisely because of cultural overlap. Here is some background on the design of the questionnaire: "The questionnaire was based on some piloting where I [Jack De Havas] got participants ($n = 16$) to free-associate their impressions and descriptions of the Kohnstamm phenomenon. . . . My approach was to be as open-ended as possible and not bias participants by using words like 'involuntary' etc. As I recall most participants had experienced the phenomenon before (as children), but not all. I then grouped the responses into themes and created a balance of confirmatory and dissenting questions" (pers. comm., August 17, 2016).[18] In this light, the similarities between scientists' questions and our ethnographic reports add reciprocal credence to the quality of both sets of descriptions. We are assured that both in and out of the lab, the experience is culturally framed.

Other components of the study—components dealing directly with the perception of weight—clearly separate themselves from low-tech, playful descriptions of the experience. Throughout the procedures, the scientists tracked instrumental measurements of electrical activity in subjects' muscles—a technique known as electromyography, or EMG—during the Kohnstamm experience. Then, via a clever setup in which subjects were asked to estimate the amount of weight they felt their involuntarily arm-lifts could hold, Jack and his colleagues found a significant mismatch between those estimated feelings and the amount of weight their lifting arms could actually have lifted, correlative to EMG levels recorded at the time of the estimation. As it turned out, subjects' estimations of the amount of weight their floating arms could hold were markedly higher—nearly two times higher—than the amount of weight the EMG levels recorded during the Floating Arms experience suggested they could hold (De Havas et al. 2016, 163).[19] Measured against the scientists' special instruments, the participants' experiences constitute what we like to call an *illusion-weight illusion*.

If we take the illusion-weight illusion described above to be a real phenomenological occurrence, then we are forced to admit that this occurrence can only "really" happen in the intersubjective contexts of the technological processes of scientific investigation. While we doubt that many folklorists would judge the illusion-weight illusion to be a folkloric tradition, we also recognize interesting parallels between the children's Floating Arms variants and the scientists' experimental designs. Performance positions, process-based instructions, priming periods, and intersubjective reactions to the experience permeate both situations. But even more central to the nature of the experience of both Floating Arms and the illusion-weight illusion is the discursive choice—which we take to be performative in both the folkloric and the scientific versions—to frame the experience as an experience of weightfulness or weightlessness. Remembering that the experiments in question were ultimately designed to study voluntary and involuntary movement, consider these references to weight in the conclusion: "We also found implicit evidence regarding the experience of involuntary movements. Participants reported that the 'floating,' stationary arm could support surprisingly high weights. . . . We used a quantitative method to assess experience of the aftercontraction based on weight-perception. Like previous qualitative studies, we also found that the aftercontraction was perceptually overestimated relative to equivalent voluntary contraction." So unlike in a previous 2015 experiment in which De Havas and his colleagues had measured the overestimation of the aftercontraction via strain gauges that measured force (compared to subjective reports of perceived force), the 2016 study chose to focus on the perception of weight.[20] Which is to say that weight only came into focus as the scientists directed the discursive elements of the experimenter-subject interactions. It is precisely this ability for verbal intercommunication to affect youth's perception of their own bodies that allows for such fantastic variations of the prototypical Floating Arms form, including play with Frankenstein-like puppeteering (D9), floating invisible bubbles (D8c), and an interaction with the devil himself (D10).

Weight and the Power of Suggestion

The range of verbal descriptions and frames for the Floating Arms experiences found in our folkloric catalog *and* in the scientists' labs speaks to the susceptibility of youths' perception of their body weight to the powers of suggestion. But where Floating Arms couples the sensation of the Kohnstamm phenomenon with relevant descriptions, other folk illusions create a peculiar feeling of heaviness without the inclusion of a recognized sensorimotor phenomenon like the Kohnstamm. A well-known example is a folk illusion we call Sandman.

Sandman (F6), like Light as a Feather, is frequently played at night during sleepovers. In a typical performance of Sandman, the patient lies with her eyes closed on the floor in a quiet, darkened space.[21] The agent sits beside the patient, possibly on the floor just above the patient's head. The agent then goes on to tell a story about the Sandman, an intruder with malicious intent who has broken into the patient's house. The story often includes the coupling of narrative elements with sensory inputs (similar to a performance of the Chills). As the Sandman stomps up the stairs, the agent pounds the floor around the patient's body. The Sandman eventually enters the patient's room and strikes the patient on the head—a narrative component that is coupled with a gentle strike on the patient's head. Then, the Sandman begins to methodically open the patient up by cutting surgery-like incisions on the patient's head, arms, torso, and legs. For the patient, the narration of these actions is coupled with haptic perceptions of the agent's fingertip miming each slice. Once sufficiently opened up, the patient is then filled with sand as the agent presses fisted knuckles into the patient's arms, legs, torso, and head. Filled with sand, the patient is prompted to pop out from the narrative frame and try to actually get up off of the floor. The patient reports a feeling of heaviness and difficulty getting up.[22]

We know of no experimental study of Sandman per se, but experimentalists have long studied the effects of suggestibility in the contexts of hypnosis. In classic hypnosis sessions, psychologists deploy ideomotor suggestions in order to induce a hypnotic state in their patients. University of Tennessee Emeritus Professor of Psychology Michael R. Nash's compelling introduction to the techniques of clinical hypnosis can be found in his co-edited *Oxford Handbook of Hypnosis* (2012). There, he lists *the arm-drop induction* as a powerful induction technique that allows the therapist to observe the motor responses of the patient. What follows is Nash's script for arm-drop induction (imagine that the speaker is attempting to hypnotize his patient):

> I would like for you to sit comfortable in your chair and close your eyes, that's right, and I would like for you to hold your right hand and arm straight out in front of you, palm facing down, that's right, hold your right arm straight out in front of you, palm facing down. I would like you to concentrate on this right hand and arm and be aware of every feeling that you are having in your right hand and arm right now at this very moment.
>
> As you know, a person is not usually aware of all the sensations he is having in his body because he is not paying attention to the particular parts of the body where these sensations are taking place. But if you pay attention to a particular part of your body as you are now paying attention to your right arm and hand, then you become aware of many different things.
>
> As I have been talking, perhaps you have noticed a feeling of tension in your hand or in your arm or maybe you have noticed a tingling sensation . . . or a tendency for your fingers to twitch ever so slightly . . . or you have noticed something I have not mentioned. . . . There may be a feeling of warmth or coolness in your hand or your arm or in both. . . . Pay very close attention, close attention to your hand and arm and tell me now, what are you experiencing?
>
> That's good. Now, I would like you to continue to pay very close attention to your arm because something interesting is about to happen to it. It is beginning to get heavy . . . heavier, and heavier and as I continue to talk you become aware that thinking about this heaviness creates a tendency for your arm to become heavy, very heavy and you will find that in a moment your right hand and arm will become so heavy like lead, so heavy that your hand and arm will gradually very slowly begin to move down, more and more down. Notice what it feels like. (2012, 491–92)

Just as was the case in the neuroscientists' labs, the hypnotist's clinic clearly shares several socially instantiated goals with folk illusions' performance contexts. Like the tale-telling agent in a performance of Sandman, hypnotists like Nash want for the phenomenological experience to be successfully realized in the patient. Even though reasons for this differ—the hypnotist's reasons being clinical and the agent's reasons being ludic—the dynamics of reflexivity remain the same. Social goals create individual embodied experiences.

Key differences between Sandman and Nash's arm-drop induction are the third-person character of the intrusive Sandman and Sandman's inclusion of haptic sensations correlative to the semantics of the narrative episodes. The former shines a light back on the difference between the two performances' ultimate goals. The hypnotist wants his patient to be calm and comfortable. Sandman's social contexts invite and create states of anxiety and fear—especially for the victimized patient lying on the floor. The haptic inputs of Sandman only add to this heightened sense of dread, so much so that the concluding experience of being weighed down by sand can be understood as suffering a weight illusion or as being literally scared stiff.

We do know of a different folk illusion that plays with the power of verbal suggestion and the perception of one's own body weight in which the agent/director never actually touches the patient/actor. We call it Buckets of Water (F5). In a performance of Buckets of Water, the actor is instructed to hold her arms out at her sides as though she is holding two buckets of water. Then, the director mimes ladling water out of and pouring water into the buckets. The actor—perceiving what we could think of as the suggestion of these mimed actions—feels her arms raising and lowering as the buckets become lighter or heavier. We learned of Buckets of Water via a remembrance we gathered from a folklore student in the spring of 2014. She did recall that she had played it somewhere around eight to ten years old, but she could not remember when she had first been exposed to the activity. She also did not mention hypnosis in her discussion of the folk illusion.

This latter omission is especially interesting because it seems that the hypnotists use the very same imaginary scenario of the arm-drop-induction technique. If a hypnotist's patient does not react by lowering her arm after Nash's script reaches the point at which the hypnotist asks his patient to "notice what it feels like" (see above), Nash explains that "the following component might be added":

> Your arm is more and more heavy and you are becoming more and more relaxed and hypnotized. Now as I continue to talk, you can imagine holding a bucket in your right hand. It is empty right now, holding a bucket in your right hand, holding a bucket out in front of you in your right hand. You can sense its

color and its weight right now. . . . What color is it? I am going to count from 1 to 10 and with each count I will silently pour a quart of water in the bucket and your hand and arm will grow heavier and heavier, moving down more and more.

> 1. I am adding a quart of water, can you picture the water right now. I have added a quart and your hand and arm are feeling much heavier.
> 2. I am pouring another quart of water in, you can hear the water pouring in the bucket and your arm getting heavier and heavier, moving down more and more. . . .
> 3. The third quart . . . (2012, 492)

The similarities between the bucket episode of the arm-drop-induction technique and Buckets of Water are, indeed, striking. And in these heavy-bucket illusions, we recognize yet another situation in which it is difficult to know whether playful children or analyzing scientists and clinical caregivers got to the activity first. No matter who got there first, it seems the illusion can be experienced in both contexts, and it may be the case that inspiration to participate in suggestion-based weight illusions flows both ways—from science to folklore and from folklore to science.

While popular cultural representations of hypnosis can conjure forth anecdotes of untrustworthy Svengali figures taking advantage of naive participants, the science of hypnotism has recently reemerged, bolstered by the advent of neuroimaging studies.[23] Using functional magnetic resonance imaging (fMRI) technologies, experimenters have been able to observe physiological changes in brain states during moments of hypnotic suggestion.[24] In 2004, Stuart Derbyshire and his co-experimenters developed a provocative experiment in which subjects' blood oxygen states in the brain were measured (via fMRI) in three separate contexts: (1) while holding a thermal device that induced physically measurable heat-related pain; (2) while experiencing the suggestion of heat-related pain in the same hand in a hypnotic state; and (3) while imagining heat-related pain in the same hand in a nonhypnotic state. Their results showed that fMRI measurements were similar during (1) and (2). On the one hand, Derbyshire and his colleagues argue that their findings "demonstrate the efficacy of suggestion following hypnotic induction in producing altered sensory experience," but, on the other hand, they remain reserved as to whether or not a hypnotic state fully explains their findings: "Nevertheless, we cannot be certain of the extent to which the hypnosis was responsible for creating the experience of pain until further studies with and without hypnotic induction are completed." They add, "Other research indicates that the hypnotic induction may be neither necessary nor sufficient to produce response to suggestion" (2004, 395, 398).

In some ways, an interpretation of Derbyshire et al.'s fMRI findings that circumvents the necessity of a hypnotized state suits our argument about folk illusions that play with suggestibility and embodied perception. We are not arguing that children and youths induce hypnosis in the actors and patients of folk illusions like Sandman and Buckets of Water—but we do not rule out the possibility either. Our argument is simply that children and youths do play with suggestibility in ways that parallel hypnosis and experimental suggestibility investigations.

In the same way that the genre of folk illusions demonstrates cultural awareness of the illusory tendencies of perception, suggestion-based folk illusions like Sandman and Buckets of Water overtly demonstrate an awareness of illusions' connections to social suggestion. Through their play with these forms, children and youths become aware of the powers that others can have over them. It is at once an embodied and social lesson. In the midst of these lessons, developed folk theories about weight eventually allow for mismatches between experience and measurability. If a child's arms feel like they are being weighed down by buckets of water, whether there are any actual buckets of water remains irrelevant to the feeling of the experience. As a matter of fact, we do not need fMRI data to come to this conclusion as its impressions are visible in folkloric traditions.

Notes

1. See Buchanan's *Japanese Proverbs and Sayings* (1965, 54). Buchanan provides this example as evidence for what he claims to be the general Japanese tendency to "delight in finding and exposing the weaknesses of others" (53).

2. See Finnegan's *Oral Literature in Africa* ([1970] 2012, 389). There, she lists the idea that mothers do not feel the weight of their own children as a variant of the anthropocentric proverb "There is no elephant burdened with its own trunk."

3. The forms in sections F and G are not the only folk illusions that play with the idea of weight. For example, Floating Arms—which we categorized with other Kohnstamm-related motor illusions—clearly plays with the perceived weight of the arms and with the idea of weight in its verbal characterization of the experience. As a matter of organization, we imagine that several different versions of our catalog could meaningfully categorize the folk illusions we have gathered in regards to specific perceptual modes (like touch, proprioception, or vision) and/or domains of perceptual experience (such as weight, automaticity, or magnetism).

4. Ultimately, we hope that our focus on weight—given its aforementioned universality and prominence in our catalog—points toward strong possibilities for eventual cross-cultural comparisons.

5. The passage refers to Richard Nisbett and Timothy Wilson's important study on introspection, "Telling More Than We Can Know: Verbal Reports on Mental Processes," in which they argue that such judgments are based on the application of a priori theories about causation rather than specific examination of the mental/cognitive processes in question (1977, 248–49).

6. See also our description of Schrempp's "Inflection of Folk" in chap. 1.

7. Eleanor Gibson's ecological approach to human development pulls heavily from J. J. Gibson's direct-perception theories. At this point, it is probably clear that we find active-perception theories in the traditions of Helmholtz and Gregory to be more suitable for the interpretation of folk illusions, so we have not fully considered the breadth and depth of J. J. Gibson's significant contributions to perception science in this book. One reason is that Gibson, as a matter of fact, rarely addressed perceptual illusions. But it is worth mentioning that both Gibson's *direct perception* and Gregory's *indirect perception* foreground the importance of action in the world. Gibson's affordances, for example, arise from perceived possibilities for action. In developmental terms, a chair is perceived differently by an adult, who can sit on it, stand on it, pick it up, or move it around, than by an infant or toddler whose afforded interactions are much more constrained. See chap. 2 in E. Gibson and Pick (2000, 14–25); see also chap. 8 in J. Gibson's *Ecological Approach to Visual Perception* ([1977] 2015).

8. See Povinelli (2012, 14) for an introduction to *f*(weight), which exemplifies a central component of Povinelli's description of the differences between human minds and nonhuman, animal minds, mainly that humans maintain and deploy higher-order, role-based processes of reasoning across disparate perceptual contexts in order to understand and behave in the world. For a brief discussion of the issues surrounding the development of *f*(weight), see Povinelli (2012, 42–48). There, he notes that parsing out what aspects of *f*(weight) are present in the developing child is no small feat, and on some interpretations of data from infant studies, *f*(weight) develops much earlier than three years old. Our survey data, however, suggest that folk illusions that play with weight tend to be performed by relatively older children, youths, and teens (see the survey data in the appendix). Thus, a more conservative timeframe for the development of *f*(weight) suits our work.

9. We admit that, for the sake of our example, we left off audio books and possible future worlds when texts float in front of our face like fleeting afterimages. Also, it is not the case that weight or *f*(weight) is the only thing one would consider when comparing the physical properties of these reading technologies. We have not mentioned, for example, physical properties such as material composition, size, or shape.

10. Flanagan et al.'s article, "Sensorimotor Prediction and Memory in Object Manipulation" (2001), which was published in the *Canadian Journal of Experimental Psychology*, is clearly meant for a scientific audience, so it is a good example of scientists bringing "folk" knowledge into their writings for each other as a rhetorical tool for introduction or juxtaposition. Even if the first-person plural "much of us" only refers to "us scientists," here, it is difficult to imagine that scientists believe scientists are the only ones aware of folkloric play with weight illusions.

11. For descriptions of the positional overshoot/empty-suitcase phenomena, see Povinelli (2012, 38–42) and Gregory (2009, 180–82).

12. By "low-tech," we mean versions that do not have built-in, sound-producing mechanisms. While our research into gag gifts of this sort are by no means exhaustive, cursory searches of Amazon.com and comparable sites like Google show a range of fake,

lightweight foam bricks, cinder blocks, and hammers. Lydia Whitt told about her father's fake brick, which he used to throw at sporting events on the television. Ray Cashman recollects an eccentric teacher who threw a foam brick at select students.

13. In chap. 1, we mentioned Snake in the Cooler, a parallel example of the relationship between technology and high-fidelity modeling.

14. For a "brief taxonomy" of weight illusions studied in laboratory settings, see Buckingham (2014, 1623–24).

15. Povinelli articulates the difficulties still remaining for scientists who look to parse the differing processes or systems of processes that are involved in the perception of weight: "But the distinction between 'sensory-motor' and 'cognitive' is a thorny issue to begin with, and to make matters more confusing, none of these are likely to be unitary constructs" (2012, 39). Even when and where we can be fairly certain that particular aspects of weight perception are, for example, cognitive, we may still be talking about a range of differing "cognitive" processes. For not all cognition is one and the same.

16. The performance that we describe in this section was performed by students and teachers at St. Cecilia whom we had worked with on numerous occasions. We have mentioned more than once the difficulties of witnessing an impromptu performance of folk illusions, and we note that we were not present on the playground that day. A teacher, however, who knew that we would be interested in the activity, recorded the performance on a cellular phone's built-in camera, and we were able to talk with her and her students after learning of it.

For a fun example of Combine Our Strengths, see the *Late Show with David Letterman* episode that originally aired on September 17, 1987 (LoloYodel 2011). Bob Weir, of the Grateful Dead, directs Letterman and two other people in a "parlor trick" as they lift Jerry Garcia higher in the air after having performed the priming period activity. The activity begins at the ninth minute of the video posted to YouTube.

17. There are several discussions of spiritual beliefs among Catholics of south Louisiana; see, for example, Rickels (1961) and Roberts (1998). It is also true that, on more than one occasion, college-age students in south Louisiana whom we approached for remembrances of levitation play refused to speak with us because they did not think it safe to dwell on spirits of the sort. While we find less folkloristic study of traditionalized beliefs in scientific and/or pseudoscientific phenomena, we can take the students' performance for their eighth-grade science teacher (not to mention the science class that teacher was teaching) as strong evidence for a set of beliefs connected to the scientific explanations for unobservable causes—of course, the easy example of weight comes to mind.

18. About the Kohnstamm effect more generally, Jack—like us—foresees a possibility for empirically grounded cross-cultural studies to help explain the experience as a whole: "I always thought it would be interesting to do a large scale study of the subjective descriptions of the Kohnstamm phenomenon, using an exhaustive list of descriptions. Then it would be possible to do a factor analysis and determine what the independent phenomenological elements were. I would be interested to know how those elements emerge developmentally and how they vary across culture. These phenomena likely shape bodily awareness, sense of agency and personality. Ultimately it would be great if some form of mechanistic model could be proposed."

19. As a setup for this portion of the experiment, subjects lifted (or pressed against) actual weights while connected to EMG, so the scientists could catalog the EMG measures correlative to supporting actual and varying weight loads. These measurements were then

compared to the EMG recordings at the time of the subjects' weight estimations (i.e., the amount of weight the subjects perceived their Kohnstamm aftercontractions could support) (De Havas et al. 2016).

20. One reason that De Havas and his colleagues changed from force to weight was that force in the 2015 study was not easily matched across Kohnstamm and non-Kohnstamm conditions. See De Havas et al. (2015, 7).

21. We gathered remembrances of Sandman performances from Louisiana, Illinois, and Indiana; Sandman also has a strong Internet presence. Several variants can be found on message boards and websites dedicated to spooky children's activities.

22. This narrativized movement of the Sandman character, creeping closer and closer to the patient, parallels narrativized movement in ghost stories categorized as Tale Type 366, such as "Who's Got My Tailypo" and "Where's My Golden Arm"—a favorite of Mark Twain's. In these tales, a malicious character has returned in order to retrieve a lost or stolen item, such as its tail or golden arm. Type 366 is often coupled with a jump-scare (Motif Z13.1—a tale-teller yells "Boo!" at an exciting point), in which the narrator yells, "You've got it!" just as the narrative reaches the point at which the encroaching villain has made its way next to the listener. For an insightful discussion of Type 366 and Motif Z13.1 in the contexts of childhood and supernatural beliefs, see Grider, "Children's Ghost Stories" (2007).

23. For a recently published review of the relevant studies, see Oakley and Halligan's *Nature Reviews Neuroscience* article, "Hypnotic Suggestion: Opportunities for Cognitive Neuroscience" (2013).

24. Magnetic resonance imaging (MRI), and especially functional magnetic resonance imaging (fMRI), technologies arose in the last decades of the twentieth century as a way to measure blood-flow changes in the brain. While fMRI studies remain provocative in neurological and cognitive sciences, humanists and social scientists should remain cautious when considering findings resulting from fMRI studies. As Logothetis warns, "fMRI is not and never will be a mind reader. . . . The fMRI signal cannot easily differentiate between function-specific processing and neural modulation, between bottom-up and top-down signals" (2008, 877).

6

FOLK ILLUSIONS AND THE
FACE IN THE MIRROR

W E ENDED THE PREVIOUS CHAPTER WITH THE ASSERTION that—in the contexts of social perception—measurability (i.e., veridicality) is rarely if ever as important as experience. It is an assertion that carries a good deal of weight for how we have tried to encapsulate illusory and nonillusory perceptions within the boundaries of ordinary social experience. This is not to suggest that measurability and veridicality mean nothing; instead, our analysis of folk illusions as a genre demonstrates that veridicality means a great deal because veridicality pushes folk illusions into the foreground by providing a necessary, stable background. The ebb and flow of veridical and illusory perceptions, moving in and out of social focus, constitute an experiential reality that is at once capable of being qualified as more or less "true" and more or less "false." Armed with this idea, certain forms of children's folklore that have traditionally been studied in the contexts of play with the supernatural can be reconsidered within the manifest "reality" of embodied perception. Excellent examples, and the subject of this chapter, are well-known play forms in which children summon the visual presentation of ghostly, supernatural entities within a mirror.

Since Janet Langlois's 1978 study of Mary Whales traditions in northern Indianapolis, Indiana, folklorists have studied mirror summonings, a ritualesque performance during which youths stand in front of a mirror in a darkened room and recite a traditionalized verbal prompt a given number of times in order to conjure or to summon a ghostly apparition. Bloody Mary and Candyman are widely recognized variants of mirror summonings that have made their way into popular culture.[1] In a prototypical performance of Bloody Mary, for example, a youth in his or her early teens enters a bathroom and turns off the lights. Then, staring into the mirror, the performer recites the name "Bloody Mary, Bloody Mary, Bloody Mary" several times. After which, the apparition of Bloody Mary appears in the mirror. In some cases, the apparition may strike out at the reciting participant.

Folklorists have interpreted these spooky traditions in a number of ways. Langlois asserted that Mary Whales mirror summonings provided girls the opportunity to invert roles of victimization while taking instigating positions by calling on a scary ghost who was once a passive victim ([1978] 1980).[2] In 1998, Alan Dundes published his psychoanalytic interpretation arguing that Bloody Mary rituals are expressions of prepubescent anxiety about female menstruation. Linda Dégh argued in her *Legend and Belief* (2001) that mirror summoning rituals primarily address social orientations toward belief and trust. More recently, Elizabeth Tucker (2005)—using a modified Jungian analysis—suggested that the ghosts that older adolescents (i.e., college students) summon and see in reflective surfaces may represent complementary versions of their preadult selves, which are at least partially full of sexuality, shadowed impulses, and suicidal thoughts. Most recently in 2014, Czech ethnographer Petr Janeček examined the global spread of these traditions via popular culture into Czech contemporary folklore.[3]

Taken together, the literature presents a strong disciplinary position on the possibly relevant symbolism and developmental nature of the traditions—especially in regard to teenage and early adult social acclimation. Tucker, for example, imagines at least two stages of social change:

> While there are parallels between preadolescents' "Bloody Mary" rituals and college students' sightings of ghosts in mirrors, the age stages are different; the images seen in mirrors are different as well. Preadolescent girls see an aggressive mother figure [e.g., Bloody Mary] who threatens to inflict pain, an early initiation into the perils of female maturity. College students, farther from the protection of home, see aspects of their adult selves that are evolving: male/female identities, sexual relationships, and social roles. Although scientific training tells adolescents to view these supernatural images as "unreal," the images may seem more intensely real at times than everyday objects. Outside rationality, the images show young seekers a new dimension of understanding. (2005, 199)

We will have more to say about development in chapter 7, but it is worth noting that Tucker's insightful analysis echoes our concern with the simultaneous experience of the possible and the impossible. Seeing the unreal vision of ghosts or spirits in the mirror, young adults learn lessons about themselves.

No folkloristic study, however, has included a detailed consideration of the primary perceptual element—that is, seeing an other, otherwise nonpresent being in a mirror. There are many reasons for this, but at least one reason is that folklorists, who work diligently not to trivialize or make value judgments about the belief systems and worldviews of those with whom we work, remain hesitant to suggest overtly scientific, materialist interpretations of reported "irrational," supernatural experiences. We also do not want to diminish or trivialize any folk illusions' attendant beliefs for precisely this reason. With the goal of deciding whether or not mirror summonings like Bloody Mary can or should be considered a folk illusion, however, we tread lightly and consider the relevant experimental literature below.

Mirrors as Phenomenally Askew Experiences

What do we see when looking into a mirror? We could say that we see the reflection of reality. But this would not be totally correct because the reflection itself is a part of reality—even if we confine our definition of reality to the realms of subjective experience. Looking into a mirror, we are reminded that our own perspective remains just one of a multitude of possible perspectives, for mirrors reflect the reality of subjectivity. Or to put it conversely, subjectivity reproduces reality in the mirror.

Philosophers and scientists have notoriously struggled with the *mirror puzzle* because images in mirrors are reversed from left to right but are not reversed from top to bottom. In *Mirrors in Mind*, a powerful book that examines scientifically and culturally the perceptual enigma that is mirrors, Richard Gregory offers his solution to the mirror puzzle as a simple perspective shift: "A mirror allows us to see the back of an opaque object though we are in front of it" (1998, 100). He describes a trick that can help prove his point. Write the word *mirror* on a transparent object like an overhead transparency (if you cannot locate an overhead transparency, try using a clear plastic bottle or a Ziploc bag). Now, stand in front of a mirror while holding the written word in front of you. You will note that the word does not flip from left to right in the mirror's reflection. How can this be?

Maybe the easiest way to solve the puzzle is to imagine yourself as not yourself—but as the person who is staring back at you. Then, imagine yourself meeting the real you face to face while walking along. Just like any other person you approach, the right hand of the you in the mirror opposes the left hand of the real you, and vice versa. The concept is easy enough to grasp,

Fig. 6.1. See anything missing in this photograph? Only with the advent of modern picture-editing software are we able to photograph ourselves approaching ourselves in the mirror. (Photo courtesy Aiden Baker-Murray Photography.)

but the puzzling thing about the pseudoperspective of the you in the mirror is that the subjectivity of mirror-you is no more real than the imagined face-to-face encounter with yourself that helps you solve the puzzle. The you in the mirror is an illusion. Returning to Gregory's clever trick, we see that the fact that the word *mirror* does not flip horizontally when you—the actual you—holds it in front of the mirror confirms that even though the mirror allows you to see the backside of objects, you are still forced to see the backside of the object from your own perspective . . . puzzling indeed!

One aspect of the mirror puzzle that we find illuminating is the fact that it frequently does not appear to be a puzzle. We are not puzzled when we step in front of the mirror every morning in order to ensure that we are presentable as we head out to work or to pick up a sack of groceries. Psychologists tell us that human children react to the visual presentation of the self in the mirror around the age of two years old.[4] By the time youths reach preadolescence, the earliest stage at which they typically report performing mirror summonings, their inner Sherlocks have already assimilated the mirror illusion into their world of everyday, very possible perceptions of themselves. The appearance of mirrors' more puzzling qualities, it seems, relies on unlikely situations like when a script is written on a transparent object . . . or when we view a mirror image in a darkened space.

Strange-Face-in-the-Mirror Illusions

One observation about the perceptual nature of mirror summonings is already clear. The disappearance into the expected milieu of the everyday of the mirror's puzzling qualities is reversed in the socially heightened moments of fearful mirror gazing that accompany the folkloric traditions like Bloody Mary and Candyman. But before we address these social frames, let us just consider what we know about looking into a mirror in darkened spaces. It is telling that in everyday use, the mirror's functional purpose—a real-time visual inspection of body image—requires that we view the mirror with the lights turned on. This must be especially true since the advent of modern electricity. Ask yourself, when was the last time you stood in front of your mirror and forgot to turn on the light?

In 1968, psychologists Luis Schwarz and Stanton Fjeld first studied the experience of gazing into a mirror in low light, which means that psychological experimentation involving mirror gazing in low light precedes folkloristic awareness of ritualesque mirror summonings by about a decade. Schwarz and Fjeld's study involved a group of sixty-four participants—three-quarters of whom had been diagnosed with psychopathologies and one quarter of whom had been designated as "normals." Subjects were placed in front of a sixteen-inch square mirror with a small recording light placed approximately three feet behind the subject. These experiments took place from 8:00 p.m. to 9:00 p.m. After a five-minute adjustment period, the subjects were asked to stare into the darkened mirror while verbally reporting their experience. Scientists working with mirror perception had known that certain subjects diagnosed with psychopathologies like schizophrenia and anorexia often report "distortions in perception, with feelings of de-personalization while mirror gazing" since at least the work of Paul Schilder (1935), so the tests' purpose was to gauge a range of phenomenal qualities accompanying such illusory visions.

To the surprise of the scientists, the control group of "normals" reported seeing and experiencing phenomenally askew visions and somatic proprioception at rates much greater than chance. One "normal" male subject reported, "My eyes are whole caves with dancing skeletons; the ladies are wearing black trousers, and the male skeletons are covered with white gauze" (Schwarz and Fjeld 1968, 279). A female "normal" subject reported,

"I feel as though the chair is rocking back and forth . . . as if it were a swing" (280). Interestingly, some types of perceptual distortions were experienced by "normals" more frequently than subjects diagnosed with certain psychopathologies, such as antisocial disorders and schizophrenic depression.

More recently, an Italian psychologist, Giovanni Caputo, gained a good deal of scientific and popular attention for his 2010 publication "The-Strange-Face-in-the-Mirror Illusion," which appeared in the flagship journal *Perception*. Caputo's study does not cite the earlier work of Schwarz and Fjeld, but his experiments—at least—seem to confirm those earlier findings. Here is Caputo's description of his study:

> I describe a visual illusion which occurs when an observer sees his/her image reflected in a dimly lit room. This illusion can be easily experienced and replicated as the details of the setting (in particular the room illumination) are not critical. The observations were made in a quiet room dimly lit with a 25w incandescent light. The lamp was placed on the floor behind the observer so that it was not visible either directly or in the mirror. A relatively large mirror (0.5m × 0.5m) was placed about 0.4m in front of the observers. . . . The task of the observer was to gaze at his/her reflected face within the mirror. Usually, after less than a minute, the observer began to perceive the strange-face illusion. (2010, 1007)

Caputo worked with fifty "naive" individuals, "naive" meaning that none of the participants reported having practiced a mirror-gazing activity quite like the one designed for the experiment. As to whether or not any of Caputo's subjects had previously played some ritualesque activity like Bloody Mary, we do not know.

It is important to note that Caputo did not come to his experimental design via personal memories of mirror summoning rituals; he only came to learn about Bloody Mary and other mirror summonings after he published his article in *Perception*. When we approached him about his compelling work, he kindly supplied the following origin story: "Actually, when I found strange-face illusions (on myself first) I was completely unaware of magical and summoning traditions. About 2004, I was studying the skill of actors to identify multiple personalities or characters in theatre. I had also built an octagon of mirrors within which a subject could sit and observe her or himself from multiple perspectives. My first experiment, unpublished, was a sort of 'summoning technique' since I asked participants to improvise monologues while gazing into the mirror. Reactions were in some cases very strong, like to weep or laugh immoderately" (pers. comm., April 3, 2015). Thinking about Giovanni's origin story alongside folkloric

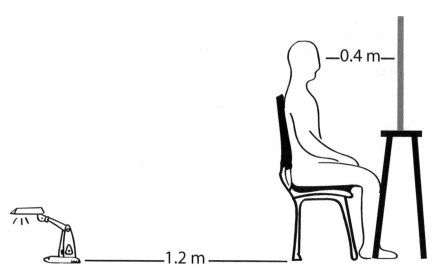

Fig. 6.2. A cross-section of Caputo's experimental setup. (Recreated with permission of the author.)

mirror summoning traditions, we find, once again, an example of a scientist creating a kind of variant social context that includes many of the folkloric tradition's components of embodied performance while producing a slanted version of the semantic frame. We wonder, What are the intrinsic parallels between gazing into a mirror in order to verbally summon forth a supernatural spirit and gazing into a mirror in order to verbally call forth some hidden, other version of oneself?

As was the case with the experimentally derived descriptions of Jack De Havas's Floating Arms experiments we addressed in the previous chapter, the descriptive reports of the visual experience provided by Caputo's subjects after a ten-minute mirror-gazing session are strikingly similar to folkloric descriptions of bloody faces and matriarchal figures summoned into mirrors:

- 66% of subjects reported a huge deformation of one's own face
- 18% reported seeing a parent's face with traits changed [of whom, 8% were still alive and 10% were deceased]
- 28% reported seeing an unknown person in the mirror
- 28% reported seeing an archetypal face, such as that of an old woman, a child, or a portrait of an ancestor
- 18% reported seeing an animal face, such as that of a cat, pig, or lion
- 48% reported seeing fantastical or monstrous beings (2010, 1007)

Another compelling similarity is the fact that Caputo's subjects found the strange face in the mirror to be the face of an other: "The participants reported that apparition of new faces in the mirror caused sensations of otherness when the new face appeared to be that of another, unknown person or strange 'other' looking at him/her from within or beyond the mirror. All fifty participants experienced some form of this dissociative identity effect, at least for some apparition of strange faces and often reported strong emotional responses in these instances" (2010, 1008). Subjects reported that the other viewed them with an "enigmatic expression." Other subjects reported that the face in the mirror gazed upon them with malice and made them anxious. The apparition of monstrous faces unsurprisingly produced fear.

As for explanations, Caputo notes that seeing our face in the mirror engages a broad "set of processes as the image duplicates one's own face perfectly in space and time, triggering an integration of perceptual, motor, and proprioceptive processes" (2010, 1008). Because the usual binding traits of face, nose, and eyes are distorted by dim lighting, the gestalt of face recognition is distorted piece by piece. And since perception never fails to categorize—even if it categorizes erroneously—an unknown face presents itself in the subject's visual field. He adds, however, that no top-down or constructivist theory of face processing can account for the visions of animal or monstrous faces. Those visions, he believes, arise from the depths.

Verbal Summoning as a Perceptual Priming Period

At this point, many of the interesting parallels between the ethnographic descriptions of the children's activity and the parameters of the scientists' experimental procedures are evident. We see in both—mirror summonings and strange-face illusions—the context of a dimly lit space, looking directly into a mirror, and the passing of some amount of time, a priming period. Can it be that the verbal components of mirror summonings which require sometimes dozens of recitations, act as a kind of self-imposed priming period and allow for the strange-face illusion to arise?

Caputo reports in his initial 2010 study that observers usually begin to experience phenomenally askew visions in the mirror in less than a minute. This timeframe is comparable to other priming periods that we have noted in folk illusions—for examples that we have not addressed explicitly in preceding chapters, see Touching Invisible Glass (A6), Furry Air (A7), String Pull (A11), Afterimage (B7), Disappear (B10), Magnetic Rocks (D2), Tied-Up Fists (D7), the Invisible Ball (D12), Falling through the Floor (G1), or Legs through the Floor (G3). Looking at descriptions of mirror summonings in the folkloristic literature, we do find some descriptions of the verbal summonings and ritually mandated time spent in front of the mirror that match the time period that Caputo identifies.

In Langlois's conversations with a twelve-year-old about summoning Mary Whales, for example, we find repeated references to the need to "call" Mary Whales up to one hundred times:

> JL [JANET LANGLOIS]: Tell exactly what you had to do to call her.
>
> G: Ya, you had to, you could call her ten or a hundred times and call her name and say, "I do believe in Mary Whales" and, a, she was supposed to come.
>
> JL: And come in the mirror? That's right?
>
> G: And come in the mirror. . . . All by myself, I called her a hundred times, and I was saying, "I do believe in you, I do believe in you." And I was by the light switch and she came and her eye started bleeding and then I cut on the light real fast. ([1978] 1980, 211)[5]

Later in the same interview, G supports the necessity of one hundred verbal calls as she describes how she first heard of the tradition from her friend Monica: "She, she tell me that. Her and her friend did it and they, she said that she *had* to call her name a hundred times" ([1978] 1980, 213; emphasis ours). Variations on the number of times a verbal summon should be repeated are inherent within these traditions. As we mentioned above, Dégh's analysis focuses on the belief professed in the verbal summonings: "She [Mary] seeks compassionate children who are willing to give her a chance to show her disfigured face in the mirror. . . . The key in this legend is believing and trusting. The daring diviner must repeat many times (from 10 to 1,000) the phrase 'I believe in Mary Worth' to persuade the unfortunate ghost to show herself. She exists because someone believes in her" (2001, 244). If we situate Dégh's observation within the model of priming periods in the genre of folk illusions, we can imagine a direct relationship between the depth of enthusiasm with which a performer believes in Mary Worth and the number of times that a child would be willing to repeat the speech act that confirms the belief. And if teens truly are priming a visual experience akin to Caputo's strange-face illusions, the lengthened time spent in front of the mirror—correlative to the time it takes to say "I believe in Mary Worth"—would be central to the experience's success.

We find another compelling discussion of the time spent in front of the mirror during a summoning in Tucker's aforementioned study of college students at Binghamton University in the state of New York. The example we have in mind came from one of Tucker's students, Christie, who slipped the following note under Tucker's office door:

> Jessica (room 207) was meditating in front of the mirror when the face shimmered and changed to the face of an unfamiliar male. Jessica, a tad frightened, left the room for a few minutes, returning with Jason (room 201), who felt a "presence" of some sort in the room. (Jason is versed in shamanism and has felt something before in 207). They went downstairs to the subbasement, determining that the "presence" was also noticeable in the far corner by the outside exit. Jason's conclusion is that the "ghost" is probably the spirit of a student here, that either lived or had extremely close friends in rooms 207 and 208. The presence in the subbasement which had malignant overtones lacking in the other locations is most likely where the event that led to the ghost's coming into being took place. Based on what he and Jessica felt, Jason concluded this was probably in the 1970s. (2005, 189–90)

In response to this note, Tucker wonders what Christie could have meant in her description of Jessica's meditation in front of the mirror: "Was Jessica just staring into her mirror, contemplating her own image, or was she trying to summon a spirit?" Can we understand this meditative action to simply mean staring into the mirror? It is difficult to say, but Tucker—like Dégh—reports a possible connection between time spent in front of the mirror and the level of enthusiasm associated with the performer's sense of belief: "Friends of Jessica described her as 'psychic,' 'sensitive,' and intrigued by the supernatural. Although none of her friends thought she had tried to raise a ghost, they were not surprised that she wanted to figure out why the ghost appeared. Her trip with Jason to the subbasement was part of a serious investigation that a student with less interest in the supernatural might not have chosen to pursue" (2005, 190).

We can think of other examples of folk illusions in which belief in the supernatural phenomenon at stake could interact with the physiological actions standard to the form. Forms that play with the Kohnstamm effect are good examples. Is it possible that children who believe in the experience of Floating Arms push harder and for longer periods of time against the inside of the doorframe?

Dundes's psychoanalytic study of Bloody Mary traditions also includes a remembrance—gathered from one of his students at the University of California, Berkeley—that highlights the importance of the performer's time spent staring into the mirror:

> During recess at school you go into the girls' bathroom. Your friends wait outside because only [one] person is allowed in at a time. One girl stands at the door to turn out the lights once you're positioned in front of the mirror. Once

the lights are out, you close your eyes and turn around three times. Then you open them and stare straight into the mirror and chant, "Bloody Mary, show your fright. Show your fright this starry night." You have to chant slowly so she has time to come from the spirit world. Then you wait to see her face. Once you see her, you have to run out of the bathroom where your friends are waiting. If you've sinned or done anything evil in your life then you will have three scratches of blood on your cheek. (1998, 123)[6]

Here, we recognize a bona fide supernatural explanation for the priming period: Bloody Mary needs time to travel from the spirit world to the girls' bathroom. Notice too that the obligation of patience is placed on the performer—not on Bloody Mary. The girl must wait to see the face.

For Dundes, Bloody Mary constitutes "a prepubescent fantasy about the somewhat fearsome but inevitable onset of menarche." In his closing section, he recounts these central elements of youths' performances of Bloody Mary: (1) It is usually enacted by an individual girl or an all-girl group. (2) It is performed in a bathroom. (3) It involves a bloody image. (4) Sometimes it involves a bloody self-image. (5) It may conclude with the flushing of a toilet. Then, he immediately adds this to his summation: "I would hope that anyone proposing an alternative theory of the Bloody Mary ritual would be able to account for all these various distinctive features" (1998, 132). It would be difficult to argue against the symbolism of blood and the bathroom, and we make no such arguments.

Whether or not psychopathological anxieties about menstruation constitute *the* core significance of the tradition, however, is less clear. Some aspects of Dundes's study also support an explanation of the tradition that, at the very least, coincides with the formal properties of the strange-face illusions. There is, of course, the importance of the priming period featured in the remembrance provided above (Dundes, by the way, never draws attention to this significant element). But there is also Dundes's own allusion to the importance of the bathroom. A bathroom space prototypically features a mirror not unlike the mirrors utilized in experimental investigations of the experience, and it is a space that can easily be made dark since bathrooms rarely feature large windows (low lighting, of course, being another central component of the performance space described in experimental designs).

Recently, we gathered a remembrance of the tradition that more clearly links the darkness of the bathroom to the vision of Bloody Mary. The following description was provided to us by a student at Indiana University in the fall of 2013. She remembers performing Bloody Mary with her friends at a slumber party, probably around when she was in sixth grade: "Bloody Mary is a game when you say 'Bloody Mary' three times in a bathroom with the lights off. After you turn on the lights, you allegedly see a bloody female in the mirror. My friend screamed before the lights were turned on because she claimed she saw a bloody figure. I left the room and didn't go pee for the rest of the night." We note that this remembrance was provided to us as an example of an activity that fits into the category of folk illusions. Taking these descriptions that key on the importance of the darkness of the bathroom and the Berkeley student's reference to the importance of a priming period together, we must admit that whatever the sociosymbolic "meaning" may be, the possibility of a very real visual experience of an entity in the mirror is supported by both the folklore and the science.

Are Mirror Summonings Folk Illusions?

As we have aligned the similarities between folkloric mirror summonings and scientists' strange-face illusions, we have tried to remain noncommittal as to whether or not mirror summonings like Bloody Mary or Mary Whales should be considered folk illusions. It is a complicated question for which we give our (still slightly noncommittal) answer now: mirror summonings are better understood if we consider them within the analytical genre of folk illusions. We hope our answer gestures toward the important similarities between the scientists' and youths' versions—especially distorted sight resulting from low lighting, the presence of a mirror, a priming period during which the visions realize. Just because these similarities exist, however, we do not argue that every performance of ritualesque mirror summonings is a folk illusion for two reasons. The first reason considers the prevalence of truncated mirror summonings, which we examine in our discussion of I Hate the Bell Witch variants. The second reason is based on the philosophical problems of ethnography—a topic we'll take up in this chapter's conclusion.

I Hate the Bell Witch, I Hate the Bell Witch, I Hate . . .

Located in the northern farmlands of middle Tennessee's Robertson County, the town of Adams is less than five miles south of the Kentucky line and some ten miles northwest of the county seat, Springfield. It is a rural place—*country*, as middle-Tennesseans put it. In 2010, the United States Census reported Adams's population at 633. But more than most small towns that

Fig. 6.3. A welcome sign off of the highway, heading north into Adams, Tennessee.

dot the hills surrounding Nashville, Adams is well known throughout the state and the region.

The Bell Witch of Adams gained notoriety in middle Tennessee when she haunted the family of John Bell during the early decades of the nineteenth century.[7] John Bell and his family migrated to the Red River area of present-day Adams, Tennessee, in 1804. In the family's first decade at their new home, the Bells found moderate success growing and selling dark-fire tobacco. But in 1817, the witch, who came to call herself Kate, started her haunting. The resulting troubles included nighttime assaults on the sleeping children whose covers were frequently pulled off, whose hair was pulled and tied to bedposts, and whose faces and arms were whipped and scratched. The Bells noticed inexplicable sounds and knocks about the house on evenings, and eventually, John Bell Senior, himself, succumbed to a portion of the witch's poisonous potion.

The Bell Witch is so well known throughout the state that a localized variant of mirror summoning, commonly referred to as I Hate the Bell Witch or as I Do Not Believe in the Bell Witch, is frequently performed by youths and teenagers in lieu of or alongside Bloody Mary traditions. Teresa Ann Bell Lockhart describes the tradition in her *Tennessee Folklore Society*

Bulletin article, "Twentieth-Century Aspects of the Bell Witch": "A person must go into a dark room and stand in front of a mirror. Then he says, 'I don't believe in the Bell Witch.' He must say this three to ten times. After he does this, the Bell Witch will appear in the mirror, and the person's face will be scratched. The scratch marks are described as being long, deep, and bleeding. They are supposed to look as if a human being with very long nails made them" ([1984] 2009, 239). In Tennessee, males perform Bell Witch mirror summonings just as frequently as females. (Indeed, Barker played a version of this activity while growing up in Sumner County, Tennessee, in the 1980s.) And for the first time, we recognize a specific, locally known witch as the summoned supernatural being. This is important, because the Bell Witch's power over young people in the state directly affects our judgment of whether or not this variant constitutes a folk illusion.

A Witch's Town

K. Brandon Barker

Heading north out of Springfield, the seat of Robertson County, twenty-first-century visitors to Adams, Tennessee, find themselves immediately greeted by a monument and semiprivate graveyard, where several descendants of the Bell family are buried. Local legends gather around the looming monument. In the early 2000s, an intoxicated driver from Springfield ran through the graveyard's iron fence in an attempt to commit suicide in a head-on collision with the monument. His truck's chassis stuck on the graves before he reached his final destination. Local retellings of this anecdote center on the power of the haunting presence: "You may die because of the Bell Witch, but only if she wants you to."

North of the monument, the repurposed old high school serves as a community center and as home to the official Bell Witch Museum. The jewel of the museum is not the paintings by local artists of Bell family members and their homestead. It is the curator, Tim Henson. In the fall of 2015, we spent time interviewing Tim, whose family moved to Robertson County in 1969 to farm dark-fired tobacco. Tim graduated from Adams's local high school, Jo Byrns, in 1971. He went away to David Lipscomb University for studies—paying for

Fig. 6.4. The Bell Family Memorial and Cemetery was constructed in 1957. Notice that the stone wall to the left of the gate is newer and a lighter color than the rest of the wall. The lighter section had to be replaced after a failed attempt by a drunken motorist to commit suicide by ramming his truck into the obelisk.

his schooling with the money he made working the tobacco fields. Tim became aware of the Bell Witch legend in 1965 when his family visited Adams before moving there permanently. Thirty years later, Tim began to take the stories more seriously, interviewing locals and searching land deeds. He is a sharp storyteller with a wealth of local knowledge and keen, intelligent eyes. He was the first to tell us about the failed suicide attempt at the Bell Family Memorial Museum. Tim refers to himself as a skeptical believer.

After assuring me that the false teeth of deceased Bell family members on display in the museum were not his idea, he lamented the attention Adams sometimes receives from outsiders. As a well-respected local historian, Tim has been interviewed more than twenty times by television and movie crews. In 2007, he helped produce *The Bell Witch Legend*, a documentary that aired on PBS, and he appears in the extras of the feature film *An American Haunting* (2005), which starred Donald Sutherland and Sissy Spacek and was loosely based on the Bell Witch legends. "They come," he says, "and get what they want out of me. Then, I never see them again. I call it the Hollywood one-night stand." In retirement, Tim earns extra money selling DVDs of the Bell Witch documentary he coproduced as well as copies of a book

Fig. 6.5. Tim Henson stands next to a nineteenth-century headstone in a graveyard of Bell family descendants.

on the Bell family's history, which he self-publishes. Yours for twenty dollars apiece. Purchase both—I did.

Tim remembers that youths were performing the I Hate the Bell Witch activity when he was in high school, and he apologized that he couldn't remember the details of the activity. As he worries about the pervasive influence that computers and phones now have over children, he wonders if the kids today still care about the "Bell Witch stuff." He admits that children today are certainly involved in sanctioned Bell Witch activities, especially those associated with the community's annual theatrical performance, *Spirit*—performed around Halloween every year. And no matter what the kids are up to, people in middle Tennessee still shudder at the thought of accidentally bringing a rock home in the treads of their tires after visiting Adams, for fear of the witch's sympathetic magic.[8] Every few months or so, Tim explains, brand-new cars break down as they drive past the Bell memorial.

We talked to several youths in Robertson County and in Sumner County (just to the south) about Bell Witch summoning traditions, and the summonings still take place. After graduating in 2014 from Jo Byrns High, Nik Norrod moved to Springfield, where he works in the Springfield Guitar

Fig. 6.6. A fine banjoist and upcoming star among old-time musicians in the areas north of Nashville, Tennessee, Nik Norrod knows well the fear that permeates performances of Bell Witch mirror summonings.

Shop, just off the square. In a personal interview at the shop in fall 2015, he told us, "No matter how silly the kids think all the Bell Witch stuff is, they still get in front of the mirror and try to call her and all." As for the illusion, however, Nik remains doubtful. "I don't think any kid is gonna stay in there long enough for that to happen. I mean if the house creaks or a dog barks—phoom! They're gone."

Nik's response is typical. Take Addy Harris, whom we talked to in the spring of 2015. Her family has lived in Sumner County for three genera-tions, and she attends White House Middle School, some forty-five miles southwest of Adams. Addy recounts that the boys at her school had per-formed I Hate the Bell Witch in the school bathroom: "They had turned off the lights, and screamed, and then they ran out with scratches on their backs." Though the boys showed Addy and her friends the scratches, Addy reported that she did not believe that they were from the witch and that she knew the boys were just playing tricks on her. Then again, when I asked Addy whether or not she had ever performed I Hate the Bell Witch, she quickly responded, "No! I'm not stupid!"

Fig. 6.7. Addy Harris poses for her back-to-school calendar custom.

Maybe it is the witch's doing, but no child in Robertson or Sumner County to whom we have talked has admitted to actually staring into the mirror him- or herself and reciting the phrase the required amount of times. So while mirror summonings remain alive, these truncated performances may be the more prevalent form of I Hate the Bell Witch traditions in middle Tennessee because the witch is widely considered a genuine and powerful supernatural force in the area. Either way, these truncated examples shine a light on a mismatch between experimental manifestations of these illusions and their folkloric cousins. Caputo's 2010 participants, for example, were not already afraid of some presence in the scientist's lab, which is to say that they did not have a demotivational force pulling them away from the mirror before an adequate priming period could be completed.

Caputo confirmed with us that he doubted mirror summonings involved the same visual illusion as those in his studies precisely because he doubts children would stare, without making intentional or unintentional rapid eye movements, for a long enough priming period. As we mentioned earlier, he was not aware of mirror summonings during the research trials for his 2010 publication, but by the time we communicated with him in 2015, he had familiarized himself with mirror summonings by watching some of the dozens of online performances posted to YouTube: "I watched on YouTube Candyman, and I found people make two errors. They either place too much light in front of the mirror or move too continuously. In my experience, subjects that move or laugh etc. in front of the mirror are using defense mechanism[s] against [the] emergence of hallucinations" (pers. comm., April 3, 2015).

If defense mechanisms of any sort frequently halt or truncate the necessary inputs of Caputo's experiments, then many mirror summonings likely do not lead to a strange-face illusion. This, of course, does not rule out the possibility that the same illusion can occur in both the experimental and folkloric contexts. We just cannot know outside of detailed examination of particular performances. At least, it is nothing new to suggest that folklorists need to know the specifics of a folkloric performance in order to correctly judge the category to which the performance belongs. It seems very unlikely, however, that all mirror summonings lead to phenomenally askew visual experiences of the sort Caputo has produced in his lab. At the same time, we must admit that even these truncated performances depend on the possibility of phenomenally askew perceptual experience—a possibility that scientists and, apparently, children recognize as absolutely real.

Roses and Thorns

Claiborne Rice and K. Brandon Barker

Dell Hymes teaches us that to understand genres we must understand "what one would need to know to recognize an instance of a . . . genre or to perform it" ([1974] 1980, 126). The metaknowledge of genre that facilitates recognition and action must be searched for systematically by the ethnographer, but it is commonly wielded by insiders in order to messy the boundaries of genres in ways both purposeful and poignant. Some folk-illusion-esque traditions, we find, seem to operate more for the sake of inducing fear or pain for targeted participants than for producing any discrete illusion. Strong examples are the torturous activities often called Rose Garden or Cherry Orchard that Clai's wife, Lydia, reminded us of very early in our project.

Rose Garden involves a familiar folk illusion performance position—the forearm out, palm up, with the agent operating on the area between wrist and inner elbow. This is the same performance position as in Where Am I Touching You? (A2). The torturous catch is that the actions are intended to induce pain, not necessarily to create an illusion. Clai recorded Lydia's recollection of the activity:

> It was actually a form of bullying. You would come up to someone and say, you want a red rose garden? You try to find somebody that hadn't heard about it yet, and I mean, who's gonna say no to a red rose garden? So the person says yes, and you say, put your arm out.
>
> You would grab their hand with I guess your nondominant hand, and do all your gardening with your dominant hand. You started by plowing some rows. You take your fingernails and rake down the arm as hard as you could between the wrist and the elbow. Then you would say, "OK, we're all plowed up, now we need to plant the seeds." And so you would go down each row and you would pinch and twist. It gets pretty painful. The next step was, "you know, it's gotta rain to water the roses." And it's actually a storm, so you get your fingers and poke them like that for a hard driving rain, and then there's thunder. [Lydia slaps her arm.]
>
> And you improvise. You can make up steps, return to steps. "Oh wait, we gotta replant this one," twist, twist, twist, or "uh oh, there's a bug blight, we have to spray this toxic chemical everywhere. Close your eyes." [Lydia slaps her arm several times while saying "bang, bang, bang, bang."]
>
> So over the course of this you do get an image on the body that looks like a garden inasmuch as there's rows with places marked that

suggest they might be plants. And, um, at that point, the person is in a lot of pain, and their brain I guess is trying to deal with their arm, and then you say, "OK, time for them to sprout and bloom and grow. Ready?" And they would of course say OK 'cause there's no way out of it at this point. You've got their hand, they've consented. If the teacher comes you know you can say, well, she told me I could. So you go back and pinch each one even harder and twist it real quick. As you're pinching each one, you make a point to pretend to pull it up several inches above the arm, and it's supposed to make the person feel like there's something growing out of their arm. And it actually does feel like that. You get to the point where, I guess the final follow-up question is, you say, "OK, now can you figure out where the roses are? Aren't you glad that they aren't in your arm any more 'cause didn't that hurt?" And you can feel something, feel pain that's outside your body.

Yeah, I would say it's a form of bullying.

In our surveys of university students, we came across other remembrances of the activity. In 2014, a student reported playing "Strawberry Patch": "First, we would ask the person to extend his or her arm and then pinch their arm saying, 'First we plant the seeds.' Then rub their arm saying, 'Then we cover them with dirt.' Then tickle their arm saying, 'Let it rain.' Then pound their arm saying, 'Thunder! Lightning!' And then say, 'Now you have a strawberry patch.' Because their arm would be pink and red." While a certain painful aftereffect can be equated with the sensation of having an actual rose garden sprout up on one's forearm (Lydia, in fact, suggested as much), we cannot dismiss the intended painful effects of the activity since most folk illusions do not cause pain.

That being said, some other "illusions" are reported with prank-like, torturous components. One of our favorites, which we learned in 2014 from an Indiana University folklore student who grew up in Hong Kong, is clearly a variant of Where Am I Touching You?: "Director will use two fingers, pretending to be the fire truck, and say, 'You are standing in the middle of your arm.' Director will tell you to yell 'red light' as close as the fire truck gets to you. When you yell 'red light,' director will hit your head and say, 'Fire truck don't stop at red light.'" A slap to the forehead shocks us. It forces awareness of our bodies and prompts us to react to our immediate surroundings. The slap and the sardonic punchline snap Fire Truck's actor out of the narrative frame of the activity. Similarly, in 2010, John Anderson, who grew up in Boise, Idaho, taught us a pop-culturally influenced forehead-slap torture that went even further to incorporate the narrative frame. He called the activity "George of the Jungle." Therein, the agent stands facing the patient, who should have his back against a wall so he can't back up. Then, with his palms facing inward in the direction of the patient's head, the agent begins to swing his arms rapidly past the patient's head. The swinging motion resembles stereotypical miming actions for an axe chop. As the agent's arms fly, he sings the intro song from the 1960s animated television show: "George, George, George of the Jungle / Strong as he could be / George, George, George of the Jungle / Watch out for that tree!" Upon uttering this final word, the agent smacks the patient (who only now realizes that the agent's flying arms were, in fact, trees) on the forehead.

For certain painful forms, such as Snake's Bite (I2) or Doctor's Shot (I1), the possibility for pain is more clearly foregrounded in the verbal and semantic components of the tradition, and in all of these forms—Rose Garden, George of the Jungle, Snake's Bite, Fire Truck, Doctor's Shot—the blends of the primed mental spaces and the physiological perceptions are important. Are they illusory? Whether or not a pain on the arm is the result of an actual snakebite or rose thorn, it hurts—snakes, thorns, and doctors be damned! The awareness of other illusions using the forearm might help the patient consent to allowing the garden to be planted, and the confusing interplay between illusion and reality that follows might assist the agent in escaping retaliation. Maybe any folk illusion in which the embodied stakes are relatively high, in which fear or pain are possible outcomes, can become a meta–folk illusion simply by failing, such that a patient who never experiences the "intended illusion" can still pass the activity along as a prank or torture, believing that to have been the intended goal all along. We could call this the paranoid school of illusion reception, members of which experience only the illusion that their friends have mostly malign intentions against them.

The Boundaries of a Genre

Folklorists also know well that findings in decontextualized environments of experimental science cannot fully explain the social, expressive manifestations of any given tradition.[9] Whether or not an illusion of the kind studied in Caputo's lab occurs in any given performance of Bloody Mary,

Mary Worth, Mary Whales, or I Hate the Bell Witch can have everything or nothing to do with whether or not that performance deals with coming-of-age anxieties, social trust, or spiritual belief. This is why we have taken the actively and socially constructed experience of folk illusions to be the time and space in which cultural ideas of perception become embodied within particular performances (and performers).

We have done our best to hold up all possibilities. We recognize that such an agnostic balance is not easy to maintain or communicate. It remains true, for example, that the embodied processes, which are always involved in human perception, do not regularly enter into folkloristic studies of the body. Conversely, the examples of children's traditionalized play with illusory processes (of great significance for the folklorist) are not necessarily important to experimental scientists. We have listened to folklorists' worries that scientistic studies of folk illusions oversimplify and look to "explain away" the sociophenomenal complexities of symbolic performance, just as we have corresponded with psychologists who refuse to see past the triviality barrier and still deem children's play unworthy of critical attention, which—they argue—should be given to identifying the so-called mechanisms that give rise to perceptual illusions.[10] As Wendy Doniger asks, so do we: "When did scholarship cease to be a collective enterprise? When did the 'uni' in 'university' come to refer to ideology?" (1998, 175). By referring to forms like Light as a Feather or Sandman or Bloody Mary as illusions (when we judge them to be illusions), we do not mean to judge their attendant beliefs as well founded or false. We reckon that the experimental science of perception and the folkloric beliefs of folk illusions are complementary—not exclusive.

In her study of the twenty-first-century belief in supernatural phenomena, folklorist Diane Goldstein makes a similar claim. Tellers of supernatural tales, she notes, frequently include narrative interludes meant to rationally judge possibilities of veridicality. Goldstein provides an apropos example of a narrative gathered from an online message board:

> Our room's window and the mirror were haunted. It so happened that one day as I stood in front of it brushing my hair (long and straight) I experienced a strange thing. Half of the image was mine; the other half seemed to belong to somebody else. The mirror is rather distorted, so I thought it was an optical illusion. But as I noticed, the other image had short curly hair, white pupil-less slightly red eyes (mine are dark brown) and a beard. I was slightly confused and turned to look at the window right behind it; no one there. As I looked closely, I moved my head from side to side. On one side was my own image,

but on the other side was something truly terrible. I screamed and ran out of the room to my mom. She comforted me and told me to come with her. I was afraid but went inside; as I peered in the mirror, there was no one in there. I dismissed the idea from my mind considering I was having hallucinations. The next day however, as I stood in front of the mirror, my whole image changed and that strange man leered at me. Again I ran out, convinced that it was not my imagination. At first mom was reluctant to believe, but after a few days our sweeper complained of a man following him around. He said, I can't do my work properly with him around me; although he was alone speaking to my mom with no one around him. I began to see that man in every mirror or glass, even the ones on the cupboard. I was scared. In an attempt to end it we removed all the mirrors from the house but that man appeared in the cupboard glass; complete man in every one. I didn't go there alone. (2007, 77)[11]

Goldstein observes that this online narrative "[includes] ordinary/extraordinary contrast and case differentiation, tries experimentation to test replicability, explores the possibility of optical illusion, and cites other witnesses." The teller, she notes, is well aware of the difficulties faced by supernatural anecdotes in the onslaught of contemporary rationalist materialism: "This exploration of embedded use of evidence is not intended to suggest the ontological reality of supernatural experience but rather to illustrate that these narratives are generally well-reasoned and more to the point, *concerned* with reason. In the narrative emphasis on evidence and rational belief, we can see that the personal supernatural experience narrative doesn't exist *in the face of* modern scientific knowledge, but in content and structure it exists *because* of modern scientific knowledge" (2007, 78). Like the illusion-weight illusion (discussed in chap. 5) that only becomes a reality in the interactions of phenomenological experience and technologically assisted scientific observation, the ghost in the mirror depends on veridical experience in which mirrors *do not* reflect absent images. The teller's inner Sherlock expects to see the reflection of one continuous face—the teller's face. Because the inner Sherlock works via comparison in this way, experience is always the amalgamation of the possible and the impossible.

Our argument is simply this. It is fruitful to think of mirror summonings as folk illusions. Doing so reveals patterns of embodiment relevant to both perception and belief. We still do not know whether or not people who have experience with mirror summoning rituals react differently to supposedly innocuous, experimental presentations of the self in the dimly lit mirror. We do not understand the interconnections of top-down mental processes like belief and the more physiologically governed processes of sensation and perception. We hope that we have shown that more and more evidence—fully in line with contemporary theories of top-down/bottom-up active perception—supports the idea that belief and perception *are* intertwined, but in no way are we saying that we know how exactly belief and perception intermingle. Despite this ignorance, the fact that children and youths believe certain experiences to be phenomenally special (i.e., supernatural) demands that theories of phenomenal qualities of perception deal with belief as a part of the experience.

To our minds, the study of embodiment and perception *requires* interdisciplinary approaches—where social scientists and scientists come to know and to use each other's data sets. For even if those data sets appear incongruent, their simultaneous consideration protects us from natural blind spots. As for adjudication of reality, we can only say that we consider illusions to be a part of the totality of reality. We accept as fact the actuality of phenomenological experience. As for whether or not phenomenally askew experiences involve supernatural powers, we remain just as uncertain as when we are asked to judge similar existential questions surrounding veridical, nonillusory experience. Those answers lie beyond the boundaries of the genre.

Notes

1. According to IMDB (The Internet Movie Database), no fewer than eight feature-length films based on the Bloody Mary ritualesque forms of play have been released in the past decade (IMDB n.d.). Bernard Rose's *Candyman* was released in 1992.

2. In 2015, a twenty-one-year-old female student at Indiana University informed me that girls in south Chicago are still performing this particular tradition.

3. For other summaries of the folkloristic studies, see Tucker (2012, 396–98; 2008, 113–16).

4. The so-called mirror self-recognition test involves two-year olds who have had a mark (a sticker or a dab of rouge) placed covertly on their face, viewing themselves in the mirror. At around eighteen to twenty-four months, babies begin to reach up and remove the mark in reaction to the visual presentation of their reflected, marked selves. For a short summary of this setup, see Gregory (1998, 7–10). For a problematizing follow-up to the standard mirror tests, see Povinelli, Landau, and Perilloux (1996). There, Povinelli and colleagues demonstrate that children, who can pass the mirror test at eighteen to twenty-four months old, fail to pass delayed self-recognition tests in which a video or photograph depicting a marked child is presented to a still-marked child. Children do pass the tape-delayed self-recognition tests around the age of three and a half to four years old.

5. Langlois reports that she conducted this interview with Gia P. at a psychic fair in Indianapolis, Indiana, in 1972.

6. This remembrance was gathered from one of Dundes's folklore students in 1996. It is the first of ten remembrances listed in his study.

7. See Hudson and McCarter's *Journal of American Folklore* article, "The Bell Witch of Tennessee and Mississippi" (1934), for an early folkloristic description of the legend.

8. A variant of other cursed rock legends, most famous being the volcano rocks cursed by the Hawaiian fire god Pele (Hammond 1995; Brunvand 1998), the cursed rocks in Adams, Tennessee, are most frequently associated with the Bell Witch Cave, a tourist destination on the north end of town.

9. Writing in the first chapter of his powerful *Anthropology and Modern Life*, Franz Boas clearly articulates the problem: "The phenomena of anatomy, physiology and psychology are amenable to an individual, non-anthropological treatment, because it seems theoretically possible to isolate the individual and to formulate the problems of the variation of form and function in such a way that the social or racial factor is apparently excluded. This is quite impossible in all basically social phenomena, such as economic life, social organization of a group, religious ideas and art" ([1928] 1986, 14).

10. Brian Sutton-Smith characterizes the impasse on the part of scientists as a kind of wrongheaded adultocentrism:

> Most social scientists of growth are caught into prediction as the measure of their science, and therefore are not particularly interested in, or tolerant of, the unpredictable waywardness of everyday child behavior and the surreptitious antitheticality of child-instigated traditions that are often the concern of the folklorist. What appears to have happened is that the scientists of human development have taken an adult-centered view of development within which they privilege the adult stages over the childhood ones. It is implicit in their writings that it is better to be at the moral stage of conscience than at the earlier stage of fear of consequences; better to have arrived at ego integrity than to be still concerned with ego autonomy; better to be capable of adult genitality than of childhood latency. The "hero" story they tell, however, is a story on behalf of adults. In its "scientific" character it does not acknowledge that this version of the classic Western "hero" tale is a "vestige" of the theory of cultural evolution long rejected within anthropology. (1999, 5–6)

11. See Goldstein's chapter "Scientific Rationalism and Supernatural Experience Narratives," in *Haunting Experiences: Ghosts in Contemporary Folklore* (2007, 60–78). She lists this narrative as a typical example of online ghost stories taken from castleofspirits.com in 2005.

7

FOLK ILLUSIONS AND
BODY ACQUISITION

THOUGH THE BOUNDARIES MAY BE FUZZY, THE CENTRAL features of
the genre are now clear. We—children, youths, and adults alike—
experience illusions. Children's folklore includes intentional performances
that give rise to illusory experiences, which is to say that folk illusions manifest socially, especially as a component of social play. Play is important; like
illusory perceptions, play reorients. At play, danger can be made safe, as in
play fighting or during a playful séance. Conversely, the safe can be made
dangerous, as when a child's living room sofa becomes an imaginary boat
stranded just north of some remote Polynesian island or when a teenager's
bathroom mirror becomes a portal between our natural and supernatural worlds. These apparent dichotomies are imperative as well. The great
play theorist of folkloristics Brian Sutton-Smith has extensively argued that
safe-versus-dangerous, trivial-versus-important, and other ambiguities of
the sort point toward the essence of play and the rhetorical, cultural frames
that surround it.[1] The dualistic qualities of illusory perception and play
with folk illusions follow suit—teetering between pleasure and pain, fear
and comfort, presence and nonpresence, self and other, and most importantly between reality and unreality.

Less clear than the mere existence of the dualities that shape the genre
are the links between age and sociophysiological development that produce the desire among children and youths to perform folk illusions. Why
do folk illusions appear in childhood and—to a large degree—disappear
in adulthood? What role might illusions play for adults and other folk
groups who do not perform folk illusions? With age, do everyday, social
illusions simply fade into the background of experience? Obviously, these
questions suggest yet another set of (always possibly false) dichotomies: the

developed versus the undeveloped, the complete versus the incomplete, the child versus the adult. Inherently comparative, the questions surrounding human development and folkloric performance remain fraught with complexities. Children can be simultaneously very much like and very much unlike adults, so those working to understand the worldviews of children must always remain attentive to the pitfalls of adult projection. Adultocentrism can be just as blinding as ethnocentrism.[2]

Twisted Hands?

Claiborne Rice

Over and over again in our study of folk illusions, we are brought back to the enigma of timing. In 2013, at the Harvest Music Festival—hosted by Yonder Mountain String Band on Mulberry Mountain near Ozark, Arkansas—I caught a rare chance to witness a spontaneous performance of Twisted Hands.[3] With between five and eight thousand people attending "Harvest Fest," the onsite tents, trailers, and campers became a small city. Between the official stage acts and the intimate jam sessions, music could be heard around the clock. The festival was widely known as a child-friendly environment, and

Fig. 7.1. On top of the mountain, the main stage at Harvest Festival.

Fig. 7.2. Erin Holden stands next to a humongous Harvest Festival goer.

during the day children of all ages could be seen running and playing among the large tents that covered secondary stages around the
mountaintop.

One afternoon, during an exhilarating performance by the band
Elephant Revival, my friend Amanda Laroche, a folklorist, directed

my attention to two children—a boy and girl, ages six and five, respectively—trying to play Twisted Hands with each other. They were trying to cross their hands and clasp them but were unable to do so. One would try but fail to figure out how to both cross the wrists and clasp the hands. The other then clasped her hands first, then twisted them left and right, trying to get the arms crossed appropriately. After a few moments, their grandmother, who had been standing by, bent over them and demonstrated the movement. She stretched out her arms and slowly rotated her wrists until her hands were back to back, thumbs down, in front of her. She crossed her wrists, clasped her hands together, and turned them inside out. Both children were clearly intrigued and were trying to clasp their hands in imitation of her, but they could not. The grandmother attempted to take them through it step by step—cross your arms like this, put your hands together like this, then twist. . . . The children were still unable to do it. They would fail at every point, attempting to twist their arms instead of crossing them, clasping the back of one hand with the palm of another, or getting as far as the hand clasp before letting go, switching their arms back around, and clasping hands normally in a prayer position. After some time the grandmother seemed to accept the impossibility and gave up even as the children continued to twist their hands in different ways, sharing their interesting geometries with each other.

The grandmother, down from Norman, Oklahoma, with the extended family for the festival, spoke of showing the children clapping games, hopscotch, and jump-rope songs and how most of the time they did not seem interested. She was gratified that they were trying to learn the Twisted Hands trick, adding, "I like to show them some of the old things, because that [pointing to the iPhone now in the boy's hand] doesn't teach them anything. They don't learn any of the maneuvers they need to get along, you know, to get along in society."

Developing Folk Illusions

We deduced early on in our research that age is an important variable in the performances of folk illusions. And it may be the case that the two children's failed attempts to perform Twisted Hands have more to do with age, or, more specifically, the physical and psychosocial level of development, than with anything else. Every performance of a folk illusion depends for its success on one or more specific physical capabilities, plus the traditionalized performance positions and/or sequenced actions.

The same, however, can be said of all folk performance. Playing any musical instrument, for example, involves certain physical preconditions, such as adequate control of lungs and diaphragm; finger, lip, or tongue dexterity; or rhythmic beating. Storytelling involves a developed language faculty, with adequate memory and recall skill. Most folk performances, though, are understood to involve a person's having developed those physical capabilities in a specialized manner. The finger dexterity of an experienced guitar player is well beyond that of the average person, as you learn when you pick up a guitar and attempt to play a sequence of chords yourself for the first time. Even our everyday storytelling competency is acquired over time with practice (Sutton-Smith 1981). One thing that makes folk illusions interesting in comparison is they usually do not involve the development of a special skill in the context of the performance. True, many of them, the coordination illusions, for example, can provoke concentration on form or even prolonged practice and improvement. But what is important for most of the forms is that the individual actor be unfamiliar with his or her body's capacity in this activity and that the folk illusion involve a set of physical abilities or responses that, though they may be unusual, need not be trained.

Floating Finger (B13) is a good example in this context. To accomplish this illusion, the actor holds two index fingers, pointing at each other but not touching, about six inches in front of her eyes, then she crosses eyes or looks off into the distance. A ghostly finger appears to be floating between her two fleshy fingers. This is a frequently reported folk illusion, and as such has surprisingly few variants. The floating finger is often characterized by players as a hot dog, sausage, or frankfurter, or simply as a finger or a third finger. The most varied element in our collection is the distance the actor is instructed to hold her fingers from her face; it ranges from an inch to arm's length, and they all work. One variant has the fingers starting at arm's length then being moved closer to the eyes, with instructions to stay focused on the starting distance of the fingers. The approaching movement brings the fingers within the zone of crossed disparity and causes the unfused images necessary to the illusion.[4] This variant exemplifies well the kind of variation we commonly see in process-based instructions. In this

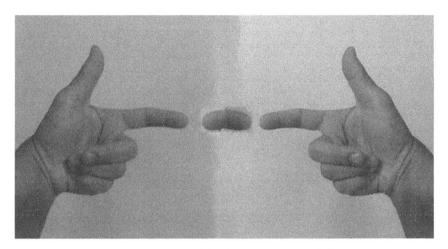

Fig. 7.3. A superimposed representation of the Floating Finger visual illusion.

case the actor can be told to look into the distance, or an additional step can be incorporated that accomplishes the same intended effect. All variants of Floating Finger involve the same few mundane physical capabilities— holding out one's fingers, binocular vision, control over focus distance with a distractor in the foreground—along with the language capacity needed to follow directions. All of these skills are present at largely adult levels by age four and a half, though some minor refinements are not completed until age eight.[5] Despite the children's early access to nearly mature visual properties, we have not found people playing or reporting to have played Floating Finger until age eight.

As our research progressed, we focused more on getting information relating folk illusion play with the age at which a person performed a particular illusion. Because the chance of observing folk illusions being performed as unprompted childhood play is so small, we used a variety of methods to gather information on the forms.[6] We talked to kids of differing ages, describing or showing them an illusion or two and asking if they know any other similar games; this was done individually or, as often as possible, in groups where kids can engage each other and forget about the investigators. We also talked to adults, asking many of the same questions, but adding questions about when and where they might have played the games. These latter remembrances obviously have shortcomings when we try to identify as precisely as possible the forms of the games and the ages at which they

Table 7.1. Folk Illusion Survey Results, Bloomington, Indiana.

Illusion	Spring 2016		Fall 2016		Spring 2017	
Total surveys completed	227		172		264	
	% Played	**Number**	**% Played**	**Number**	**% Played**	**Number**
The Chills	68.3%	155	69.2%	119	71.6%	189
Floating Arms	55.1%	125	49.4%	85	59.5%	157
Twisted Hands	44.9%	102	33.7%	58	54.2%	143
Light as a Feather	18.5%	42	15.1%	26	17.8%	47
None of the four	12.3%	28	11.6%	20	9.9%	26
No folk illusions reported	7.9%	18	7.0%	12	7.2%	19

were played, because people vary greatly in the quality of their recall, especially of an activity that even at the time was likely seen as of little value.

Fortunately both of us teach at large universities, so we can ask our classes. Brandon teaches large sections of Introduction to Folklore at Indiana University. During class he introduced the four prototypical folk illusions (see chap. 2), then asked students to complete an online survey through SurveyMonkey or Qualtrics. The surveys inquired specifically about the students' familiarity with the four illusions, including where and when they had been performed, and solicited descriptions of additional forms they might know about or have played. Table 7.1 summarizes the results from three semesters: the spring and fall of 2016 and spring 2017.

In each survey, no more than 13 percent of people reported having played none of the four example illusions, while no more than 8 percent reported performing no illusions at all. Of survey respondents who reported the ages at which they performed different illusions, the median age was about nine years old for all folk illusions. A few people reported playing at ages five and six, but the number climbs higher at seven and above. Aside from the rare exception, the college student survey data allow for a maximum age report of about twenty years old, so our information on adult performance derives from interviews and observations only.

Looking at all of our research together—interviews, observations, and survey results—we have identified three broad periods at which folk illusions are performed. The periods are identified primarily by the kind of performance, and also by the participant roles available to the performers and by the purpose for the performances.

Table 7.2. Three periods of folk illusion performance.

Ages	Typical Games	Additional Characteristics
Young child (2–5 yrs)	Got Your Nose, Kiss Your Elbow, Detachable Thumb	Adult or older child is instigator; illusions targeted at immature body perception
Childhood through adolescence (8–18)	Pat and Rub, the Chills, Floating Arms, Twisted Hands, Light as a Feather	Simpler forms earlier; more socially complex forms later
Adulthood (18–80)	Who's Touching You, Falling through the Floor, or Legs through the Floor	Illusions played as pranks, while stoned or high, or for nostalgia

Young children of toddler age up to about age five play several folk illusions as patients. We sometimes refer to these activities as the Uncle Tricks due to the stereotype of the tricky uncle who likes to tease nieces and nephews with little pranks and tricks. People frequently remember Detachable Thumb (B2), Can You Kiss Your Elbow? (E9), and Got Your Nose (E1), but their memories usually consist of playing the agent role with younger siblings or cousins, or of observing when a third person plays with a younger kid. A handful of people have reported, though, that they remember an uncle or their dad showing them how their thumb can come off. At a preschool in Bloomington, Indiana, we observed a three-year-old touch his face as if to check for the presence of his nose several times after his teacher had playfully pulled his nose off. All of the folk illusions common to this age warrant additional research. The age bands at which the illusions are effective are narrower than with older children because they target a relatively specific confluence of social and physical developmental characteristics. For example, from our observations we have found that children younger than two years old often cannot identify their elbow. Between two and three they will normally assert that they can kiss their elbow, but they usually kiss a part of their forearm or their upper arm near the shoulder. Only after age three, but continuing even up to age ten, will kids both try to kiss their elbow and realize that they cannot.[7]

The preteen and teen years encompass the central ages for playing folk illusions. The social interaction that becomes newly interesting to kids beginning around age ten and continuing through the secondary school years is integral to their interest in folk illusions. The Chills is generally reported early, with a few people reporting this folk illusion as young as age six, and many more at age eight. Unlike Twisted Hands, which presents a relatively demanding physical complexity for the actor, the Chills requires very little action on the part of the patient, so it can be performed early. We have many reports, some quite detailed, of people learning folk illusions from older brothers or sisters who learned them at school, and, conversely, reports from people who learned a form at school and taught it to their younger siblings. Light as a Feather is a form reported relatively later, at ages correlated with middle school (approximately eleven to thirteen years old), but that also is sometimes reported by people who play at seven or eight as the patient. On the whole, a handful of adult informants report quite specifically playing a folk illusion at ages six and seven. The number of reports grows at age eight, or third grade, and the most usual response to "when did you play this?" invokes a preteen year (ten to twelve) or "middle school."

The enthusiasm of people of all ages to demonstrate folk illusions for us has been one of the most fun elements of this project. When adults in an interview remember performing the illusions as kids, they are often inspired to play them once more, as if they had suddenly become twelve years old again. Both of us have seen adults at cocktail parties begin pressing their arms against a handy door frame or suddenly throwing their hands into a clasp to see who might notice their missing finger. In these situations, adults seem to enjoy the forms and can be as interested as any child to learn one that they have never seen before. But it is also true that without the provocation of the research question, such events are rare. As fascinating as folk illusions are to most adults, they simply do not play them in the same spontaneous way that kids do.

This does not mean that folk illusions are never performed by adults, and when they do spontaneously occur, three different kinds of motivations are expressed. First, the surveys netted several examples of what we might think of as the adolescent forms being extended into adulthood for a variety of local reasons, such as social rituals or traditions. One college student stated that he still played invisible basketball "whenever we are at the court without a ball," and another young woman said her family performs together an illusion her dad had taught them all "at least once every summer." Other forms that adults report playing, or that we have observed, are folk illusions that adapt readily to pranking, like Who's Touching You? (A12). A final third group that reports playing folk illusions as adults are people who have played when they have smoked marijuana or taken other hallucinogenic drugs.

Fig. 7.4. Beka having her picture taken yet again at Festival International in Lafayette, Louisiana, 2013.

One informant, Beka Riley, is a flow artist who performs hula hoops, fire spinning, and other flow arts frequently at outdoor music festivals. She described playing Falling through the Floor (G1) in her mid-twenties at a music festival while on a hallucinogen. She had never heard of it before when a girl told her she could make her feel as if she were flying. As her arms were lowered to the ground, she felt strongly like she was falling into the ground. Like many of our informants, both kids and adults, she said that she had never thought about it again until just now. "I never thought to try it not on drugs," she said. She also named several common folk illusions as occasionally being played at festivals in a similar context, such as Floating Arms and Dead Man's Hand. For some people it was a chance to experience something "freaky," while others enjoyed figuring out the mysteries of the illusions.

Not unexpectedly, people who are high on drugs are easy targets for pranks. Jaron Nazaroff, a twenty-four-year-old dermatology clinic assistant from Santa Clara, California, described some of his favorite pranks to play in late high school and college on "people who are getting stoned for the first time or who had not smoked much before. Just to play with them." One form they called Heavy Ball involved a group of friends tossing a ball around, maybe a volleyball or a tennis ball, whatever is to hand. As they toss it, they begin to observe that the ball is getting oddly heavier. They comment on it as they continue to toss and gradually begin to act as if it is heavier, exaggerating their motions to display the weight of the ball. Finally, as they prepare to throw the ball to the naive participant, they warn him that it is now "as heavy as a medicine ball" and to prepare to catch it. When the person catches it, he is all set to deal with the weight and often flips the ball in the air, staggers backwards, or even falls down, to the delight of everyone else, of course.

His favorite, though, was "putting someone in a box." This prank also involves the collusion of a group of friends. It begins when they ask permission of the naive person to put him in a box. When he agrees, they begin to assemble a cube around the person as if out of glass or clear plastic. They pretend to seal the sides with tape or glue, then as the last side is attached, the group begins to communicate silently. They mouth words with no sounds, laugh, sneeze, stomp on the floor, or drum on a chair but with no noise emitted. After a few moments the person will begin to get suspicious that he is really in a box. Everyone acts as if they cannot hear the person talking or asking questions, and carries on as if he isn't there. Eventually, Jaron said, the person might panic as he begins to doubt his senses. Jaron observed about himself, "You can tell that I'm an older brother."

While this initial study has identified at least the outlines of the ages for which particular folk illusions might occur, the problems of development and their relationship to the performance of folk illusions will ultimately need to be addressed via targeted ethnographic observations, such as in a school setting over a year, or by studies that focus on a specific form or group of folk illusions. No doubt, such studies will identify finer age grading, more closely correlated with the particular physical capabilities engaged by each illusion.

Folk Illusions and Body Acquisition

To think about the way folk illusions introduce participants to characteristics of their bodies that they were previously unaware of, we adopted the metaphor of *body acquisition*, analogous to the common notion of language acquisition. In the scientific literature of language development, the term *acquisition* is common; language is ordinarily understood as an ability people initially do not have, then develop over time, until they arrive at a point at which we can say they speak a language. The process obviously consists of progressively adding small, local but interrelated abilities that eventually work together into an ever larger whole that we conceptualize finally as a single capacity. Our body is hardly different, despite the oddness of the body acquisition metaphor. After all, do we not have a body when we are born (or before)? We would not exist without a body, would we? So it makes little sense to talk about its "acquisition." But on analogy with a Chomskyan view of language, which posits an underlying mental faculty that is gradually refined into the adult performance capabilities of a particular language, our infant bodies are primordial, underdeveloped for accomplishing any task that a human needs to do to survive.[8] Breathing, for example, is an ability evinced suddenly upon exiting the airless confines of the womb, so there is some justification for saying that the child has acquired the ability to breathe. Prenatal infants born unable to breathe must be sustained on breathing machines, sometimes for weeks, before they are able to breathe on their own. The muscular and neural systems underlying this capacity are in some abstract sense present, yet they must grow into their proper configuration, with enough strength and organization to accomplish their intended task. Once the child breathes on its own, we sigh in relief, knowing the child has taken an essential step in the adventure of life.

In an important way, then, every behavior that we demonstrate, every motion or skill that rides upon our body, is acquired. To understand an action in a human context, it makes more sense to ask specifically how any given skill is acquired rather than to argue about its innateness.[9] Some abilities will need less development of certain physical properties, less practice, and some will require more. Natural language requires a great deal of use in order to develop properly. Children born deaf and/or mute begin babble-signing at the same age that hearing infants begin to practice phoneme-shaped sounds, and if left to their own devices, such as with deaf children of hearing parents who wish not to introduce their children to signing, will begin to create signs for interacting with other people with a regularity that tends to follow a fairly predictable morphological and syntactic structure (Petitto and Marentette 1991; Goldin-Meadow 2003). If these children are not socialized with other people motivated enough to respond to and develop their home signs, they will fall behind in intellectual development. Without rich interaction with other language users, no child can develop the full range of abilities we associate with normal human language use.

Phenomenology, with its close attention to consciousness and the body, presents an effective framework in which to think about body acquisition. Maurice Merleau-Ponty describes the process of incorporating everyday habits into the body's capacity as "sedimentation" (1962, 131). Applying Merleau-Ponty's ideas, Hubert Dreyfus (1996) develops a model of skill acquisition that includes five stages: novice, advanced beginner, competence, proficiency, and expertise. Skill acquisition begins in explicit thematization of the body, but with practice, the body begins to fade from conscious attention as the skill can be performed with greater dexterity. It is this absent or invisible body that many philosophers take to be the default state of consciousness.[10] Drew Leder (1990, 83–92) coins the term *dys-appearance* (*dys-* from the Greek meaning bad, hard, or ill) to describe how unpleasant experiences like pain, injury, or shame can provoke explicit thematization of the body in one's attention. He points out that normal physiology when reaching functional limits can also cause *dys*-appearance, such as when hunger, fatigue, or dizziness prompts our body to seize our attention.

Kristin Zeiler (2010), however, argues that *dys*-appearance has an opposite, "*eu*-appearance," when the body appears in the zone of attention as good, easy, or well. She presents examples of physical exercise, sexual pleasure, and some cases of wanted pregnancies as times when a person might take notice of a pleasant "bodily feel." She argues that when a person is immersed in a pleasant experience of an exercise like swimming or dancing, for example, "the subject may attend to the body and its functions without this implying that the body is experienced as other than the self, a part of the outside world" (339). Whereas the usual philosophical notion of the invisible body posits that turning attention to the body results in the faltering of the attempt to complete a task skillfully, Zeiler insists that expertise can allow a nondisturbing thematization of the body.

Phenomenological investigation of skill acquisition has focused on the experiences of adults. Dreyfus's model of skill acquisition refers to explicitly learned adult skills, such as driving a car or playing tennis, and Zeiler examines adult habits such as swimming for exercise or conditions like pregnancy. Merleau-Ponty's "sedimentation," though, encompasses more than just explicitly learned skills.[11] Most of what is learned throughout childhood never reaches the point of thematizing the body for the subject as attention continually projects into action. But if folk illusions prompt the *eu*-appearance of the body momentarily from out of the background noise of social interaction, then the memory of the performance is layered both with the body that is thematized and the pleasure of its appearance. This *eu*-appearance perhaps accounts for the joy adults take at recollecting their experiences with folk illusions as kids.

When we examine folk illusions as a genre, two points regarding body acquisition jump out at us. First, if folk illusions are a guide, every element of body acquisition is socially implicated, and, second, not everything that develops has a self-evidently important role in life or survival. This latter point was proposed in evolutionary biology by Gould and Lewontin in 1979. They introduced the term *spandrel* to describe an architectural by-product induced by adaptive change in a complex and integrated system. Any morphological property may interact with other parts of the whole organism such that secondary developments or emergent properties arise to become part of an organism's overall phenotype, but neither contribute to, nor detract from, fitness (see also Gould 1997). Most, if not all, folk illusions reveal one or more such hidden capacities to a person—facts or capacities of the body about which the person was heretofore ignorant. Also in most cases, if not all, these capacities are trivial elements of daily life, the knowledge of which one could safely survive without. Yet we notice that children have traditionalized so many of them. Why?

In our experience, the appearance of folk illusions coincides with points of human development in which body acquisition is particularly marked. Folk illusions themselves are social through and through. The analogy with language is instructive, as Mikhail Bakhtin describes:

> As a living, socio-ideological concrete thing, as heteroglot opinion, language, for the individual consciousness, lies on the borderline between oneself and the other. . . . The word in language is half someone else's. It becomes one's "own" only when the speaker populates it with his own intentions, his own accent, when he appropriates the word, adapting it to his own semantic and expressive intention. Prior to this moment of appropriation, the word does not exist in a neutral and impersonal language . . . but rather it exists in other people's mouths, in other people's contexts, serving other people's intentions; it is from there that one must take the word, and make it one's own. (1992, 294)

Words occupy both the social and the private spheres at the same time, but the social precedes. A person takes words and wrestles them into the private and personal, though they remain Janus-faced even then.

This cooptation would seem more necessary with words than with our bodies. Words are things that we get from others, there is no denying. If we coin our own words, to have any exchange value, they must be situated in meaningful discourse. They must be reproduced by us faithfully enough, and so every word we use must be a copy of something found. Surely this is a poor analogy with our bodies. My body is not a faithful copy of anything—it is what it is, and I must settle for it. But the extension of our body is not just a matter of limbs in space. Our direction, perception, and understanding of our body is mediated by the invisible body schema that has a role in every action we accomplish. The body schema does not grow in isolation, but develops as we move and act every day, especially as we act socially.

Though we still do not understand the full extent to which our body schemas rely on social interaction for development, we do know that body schema maturation responds to language, to the gaze of others, and especially to the technologies that we use every day. When we think about body acquisition, then, we think not just about the growth or appearance of a body part, but of a motor skill or accomplishment that results from the carefully timed interaction of brain, nerves, muscles, technology, and emotions. As a package, many of these skills have common appellations— *throwing, walking, skating, typing.* Many others, however, are not so named in everyday language because they tend to be invisible, to be part of the processes of perception or action rather than an identifiable skill or behavior.

Let us reconsider Floating Finger. The ability to focus on a distant object with a detractor in the foreground is a good example of such a skill. The word *skill* even seems odd to apply to it. It could prove useful or necessary for survival, and it is appropriately learned at some point in development. We would hardly call this skill "acquired" in any real sense until we have used it in a skillful manner, such as in a game like Floating Finger. By focusing attention on one's ability to focus, the game has the effect of adding to one's implicit understanding of what the body can do. In this sense, we say that the body has acquired the capacity to selectively focus not on an object that is in the central foveal field, but at a point beyond it in the distance. So with skills that we normally recognize as learned or practiced, like guitar playing or long jumping, we say that a person has acquired a skill. Looking at skills that we do not generally name or recognize as skills per se, we tend to say that a person has matured or grown up, or achieved adult competency.

Stephen Macknik and Susana Martinez-Conde (2010) provide some wonderful examples of body acquisition (though they do not use this term, of course). They describe how children are less susceptible to adult-style magic tricks that depend on our mature senses of expectation and assumption. In the disappearing coin-toss trick, for example, the magician tosses a coin up in the air with his right hand and catches it three times, then tosses the coin into his left hand and snaps it shut. When the magician opens his left hand, viewers see the coin has disappeared. Macknik and Martinez-Conde point out that this trick relies for success in part on adult levels of attentional focus. When the magician's right hand moves to toss the coin, adult observers cannily—if unconsciously—predict the trajectory of the coin based on the hand's motions. The closing of the left hand, timed to coincide with the landing of the coin, confirms the trajectory, and a flying coin is inferred, though the coin actually was never tossed at all. According to the magicians in Macknik and Martinez-Conde's research group, this is the kind of trick they would never perform with children because the magicians know that children younger than about the age of five have not yet developed these strong attentional biases that the tricks exploit.

The researchers asked a well-known children's magician, Silly Billy (David Kaye) what tricks he has found work well with kids younger than five. He described the coin from the ear, needle through a balloon, pouring water into a cup or a newspaper cone only to have the water disappear, all of which work with younger notions of object permanence, good continuation, or with objects like balloons whose physics the children are familiar with (2010, 155–59). Certainly each child has a body, as we say conventionally, but the child's body perceives and understands in its own characteristic way. In a very real sense it has not become an adult body yet. Once it has learned *not* to attend to certain realities in the environment, but instead has been led by the inner Sherlock into predicting and anticipating the appropriate events, we can say that the person has matured.

Macknik and Martinez-Conde argue that many of the illusions unavailable to children under age five require a theory of mind to understand. Magicians often use their own eyes to manipulate an observer's gaze in their tricks, focusing, for example, on the predicted position of the coin. Gaze following is one skill contributing to the ability for joint attention, a cornerstone of "theory of mind" development, and it usually arises in human infants between eleven and fourteen months old (Tomasello 1999, 65). Children are slow to develop adult levels of sensitivity to the geometry of eye gazes, however; six-year-old children reported that someone was gazing directly at them across fixation positions ten to thirty centimeters to the left or right of the bridge of the participant's nose, with adult levels of sensitivity (about eight centimeters left or right, or about the width of one's eye) not reached until age eight or nine (Vida and Maurer 2012). The geometry of triadic gaze (two persons focusing on a third object) is even more delayed, with adultlike levels of accuracy not reached until age ten. Thus, it is not surprising that magicians find that illusions depending on misdirection by gaze or prediction of object movement are not effective on children below age five.

We cannot help but notice that the age of five also marks a break in our periods of folk illusion play. If the core of every folk illusion performance is a person experiencing an illusion, a great deal of the fun comes from knowing what the patient is experiencing. In a deep, intersubjective way, each folk illusion is experienced not just by the patient but indeed by all participants. This effect could not happen without a theory of mind, the ability we have to imagine what another person is experiencing. Folk illusions teach us that people have ideas about what the body can and cannot do. This notion is strongly reinforced for the director, the actors, and even the observers whenever a folk illusion is introduced and actually succeeds on the actor's body.

We might go so far as to suggest that people have a theory of body, in analogy to the theory of mind. People in discourse build representations of interlocutors' minds to facilitate communication. We keep in mind what other people know, not just in general but specifically in relation to the current situation and ongoing conversation. The theory of body would say that we are aware that other people have bodies, and further that other people have relationships to their bodies: They feel hot when they have a fever. They feel tired when they do not sleep. It hurts when they sprain their ankle. Folk illusions interact with this theory of body. When we see someone experience the Floating Arms effect, we identify with their surprise. If we have had the experience, we can share their response intersubjectively. The director, actor, and any observers are engaged in joint attention: They are focusing on the same imagined event, though it is not the usual example of a third object. The attention of other people creates a sort of object out of the feeling (or the feeling of the experience).

This is an important element of body acquisition. We usually imagine that the body we experience is ours, that it somehow preexists us and belongs to us privately, but so much of body acquisition happens in a social environment. When we experience something new regarding our body in a private moment, we cannot say we understand it until we have integrated it into our larger body schema that is itself mostly constructed through existing cultural categories that we have acquired through language or using other sociocultural frames. This is true in a nominal way when we learn, for example, the names of body parts, as evidenced by the fact that some languages name particular parts and other languages do not, or reference with a single word what English recognizes as both *arm* and *hand*.[12] The body schema's incorporation of cultural categories also happens when we use body parts or actions as indexes of internal states, such as inferring someone is tired when they yawn or when they have dark circles under their eyes. And it occurs when we use body parts or experiences metaphorically (and metonymically), such as someone asking you to *give them a hand* or telling you to *face your problems*.[13] In light of all this, even if we continue to try and separate language and the body into two discrete systems that interact and inform each other, in analyzing performance we need to continue to acknowledge that we are dealing with a whole system being cocreated by physiology and by sociocultural experiences.

Peekaboo

Body acquisition must be affected by the totality of embodied experience. Folk illusions, which seem to be performed in developmental bands and in fleeting contexts, do not represent great swaths of the sum total of experience. They do, however, exemplify socialization of the body as an element of expressive culture. Using folk illusions as an example, then, we want to extend our framework here to finally consider another well-known play form that contributes to embodiment in traditional and communal ways. Adult-infant play usually called Peekaboo in American English presents a primary illustration of the thoroughly social nature of body acquisition.

Peekaboo has been studied by developmental psychologists since 1929 and is well enough understood to be used in assessments of infant development and as a paradigm for research into infant psychology.[14] Peekaboo, or games very similar to it, is played around the world in a great many diverse cultures. Anne Fernald and Daniela K. O'Neill's excellent summary documents examples played in Japanese, Korean, Xhosa, Malaysian, Greek, Hindi, Persian, Russian, and Tamil, in addition to German, French, Italian, Portuguese, and both British and American English (1993, 261–67). Here is their analysis of the commonalities and differences among the Peekaboo play styles of different cultures: "The peekaboo game is effective because it exploits perceptual, attentional, and affective predispositions of the young infant, and because it both engages and accommodates the developing cognitive capabilities of the child" (1993, 268).

We have argued as much with regard to all folk illusions at the various ages at which they are played. As predispositions change with maturation, so do the types of folk illusions that garner children's interest. Once people reach an adult level of maturity, their interest fades except when folk illusions might be performed as a prank, as an activity to facilitate the recreational use of hallucinogenic substances, or as a demonstration when an adult is reminded of folk illusions by their children or by meddling folklorists. Given this natural progression of interest in the genre, Peekaboo joins folk illusions such as Got Your Nose and Detachable Thumb as a form in which it is typical for an older, more developed person to play with and try to trick a younger—sometimes much younger—playmate. These forms are representative of developmental scaffolds and social reinforcements of tradition that contribute to body acquisition.

Over the course of infant development, Peekaboo play exhibits a predictable sequence of forms. From three to five months caregivers play a "looming" style performance, where the adult is in front of the child in full view, then suddenly advances his or her face closer to the child's, looming over her with a smile before drawing back again for another round of the game. In this early variation, a vocalization like "Ahhhh Boo" will accompany the rapid approach of the caretaker's smiling face, usually starting on a high pitch and then falling off. The caretaker is rewarded with smiles, laughter, or cooing or gurgling from the infant, which participants report interpreting as signs of the child's enjoyment. The child's delight naturally prompts repetition of the game.

The next form, Peekaboo proper, occurs from five to eight months and constitutes the first of two codeveloping forms.[15] Peekaboo is usually played in *rounds*—that is, a given performance will consist of several rounds of hiding and revealing.[16] Fernald and O'Neill describe the prototypical round of peekaboo as consisting of three steps: First is the establishment of mutual attention, next is some form of hiding, and finally comes uncovering/reappearance (Fernald and O'Neill 1993, 270; Bruner and Sherwood 1976; Greenfield 1972; Ratner and Bruner 1978).

The earliest form of Peekaboo proper is recognizably adult-initiated. Seizing an opportune moment, perhaps during bathing or feeding time, the adult will cover the child's head or face with a rag or bib, then give the "alert" call, which signals the onset of the game. In American English this would be something like "Where's Baby?" with a rising intonation. Mutual gaze between the adult and child, however, is a consistent feature of the dyadic interaction, so it is rarely explicitly marked. In other words, when the adult notices the gaze is disrupted, she can initiate Peekaboo play by calling out to the child, who then fails in an attempt to locate the adult visually. The absence of eye contact becomes the salient feature of the child's environment.

5a. Adult agent highlights performance space and time with eye contact / alert call
5b. Agent initiates hiding activity
5c. Priming period
5d. Agent reappears / release call
5e. Eye contact is reestablished
5f. Agent and patient respond to each other
5g. Optional reinitiation of hiding activity

In folk-illusion terms, mutual gaze (5a) in the first round is not marked as a performance position until the second performance position (5b) is taken. At that point, the keying elements of Peekaboo morphology are in place and participants can take their roles. When the moment is right—that is, once the adult knows the child is seeking to reestablish a mutual gaze but before the child panics or frightens—the adult will remove the covering and reestablish the mutual gaze while giving the "release" call, which begins on a high intonation and falls rapidly: "There she is." It is common to see the two calls run together, with the alert call lengthened until the appropriate moment for reappearance, as in "Peeeeeek-aaaaa-Boo!"—where the first two elongated syllables correlate with the hiding phase and "Boo" with the caretaker's sudden reappearance. Fernald and O'Neill document different styles of alert and release calls, both within and across cultures, but the game's two moves, with their distinctive timing and abrupt intonation shift coinciding with reappearance, characterize the style of play around the world (1993, 263–67).

On the reveal, eye contact is reestablished, the baby will smile or laugh appreciatively, and the caretaker responds appropriately with smiles, laughs, and/or vocalizations. Once the mutual gaze is reestablished, the structural description for step 5a is also met, so the final performance position, 5f, naturally forms the first step in another round of the game. The adult, as agent, can choose to continue with the next round by then recovering the baby's face or rehiding while giving the alert call. In this way the game becomes a self-perpetuating machine, the end state of one round being the initial condition of the next, an aesthetic quality found in many art forms with repeating formal qualities, especially music and lyric poems. At some point, of course, this machine breaks down, as the attention of one participant wanders perhaps or is drawn by a more compelling stimulus, such as the approach of another person. Nonetheless, this self-perpetuating form draws both child and adult into exercising their mutual delight.

One source of many adults' amusement with Peekaboo play is that the adult believes the child thinks the adult has disappeared. Some of our interviewees expressed a reluctance to cover the head or eyes of children because doing so might scare them. Hannah Ritorto, a graduate student of creative writing at the University of Louisiana at Lafayette, mentioned a niece, Annalie, with whom she played regularly. Annalie could not play Peekaboo because she would cry when her head or face was covered.[17] In cases like this, adults may prefer instead to cover their own eyes or face, or to move their head to a position outside of the child's view. "The kids think that you are gone when your eyes are covered," said Hannah. She used the term *object permanence* to explain that the young child thinks that not seeing something means that it no longer exists. Nearly every informant we have spoken to has stated this basic idea in one way or another. We understand this to be the basic illusion operating in Peekaboo. Just as the director of Concentration (A3a) knows that no egg has really cracked on the actor's head and no knife has really stabbed her in the back, so the adult participant in Peekaboo knows that she has not really disappeared, but only manipulated conditions so that the infant perceives the world this way. This conception on the part of the adult participant aligns Peekaboo with other folk illusions played with younger children, in that they target the child's immature schemas.

Peekaboo's Invisible Agent

Claiborne Rice

At Surf'n'Turf 2017, an ultimate Frisbee tournament in Destin, Florida, I met Janna and David H., who had come over from New Orleans to play with our Lafayette team and brought their baby, Finley, along for the weekend. This tournament is always fun, with one day

Fig. 7.5. Baby Finley with her father, David, at Destin Beach.

of matches on the grass of a local park and the second day of matches on the beach. Beach day is festive, competition often taking second place to the delights of the beach setting. As Baby Finley enjoyed the sand and water with her parents, I chatted with them about games they played with Finley. On Saturday at the field, twelve-month-old Finley had taken her first unaided steps, so it was an exciting time to see them interact as a family. When the conversation came around to Peekaboo, Janna described what she thought was the earliest form of Peekaboo she played with their daughter. Early on, Janna said, when Finley was three or four months, she would lay Finley on the table to change her diaper, holding Finley's feet with one hand and raising her legs up. In that position the feet naturally block Finley's eyes, so she would say "Peeeek-a-" then lower the feet down and say "Boo!" Finley would smile and laugh, and this would keep her occupied during changing. Janna was holding Finley at this point in the interview, so she volunteered to demonstrate for us. She laid Finley on her back on an orange towel in the sand. She raised Finley's legs like she was changing her but with two hands, one on each foot. When the feet blocked Finley's line of sight, Janna gave the alert call, then lowered the feet to say "Boo," and Finley really loved it. She laughed right away, and I could tell it was a familiar, comfortable position for her. Janna again raised Finley's legs and called, but this time brought the feet apart for the reveal rather than down. They did two more rounds. In between, Janna kissed Finley's little foot and faked like she was eating or gnawing on it. What fun!

Janna and David describe that they can tell when Finley wants to play Peekaboo now because she covers her own head with a sheet or towel. The adults will respond with "Where's Finley?" and Finley pulls the sheet off of her head, laughs and looks around, then recovers her head. David states that one place Finley likes to play is in an overstuffed chair in their living room. Finley, at this point unable to walk unaided, loves to stand behind the ottoman then duck down behind it. David will call out "Where's Finley?" after which Finley will pull herself up into sight again, and David gives a surprised expression and says "There she is!" Janna and David agree that she will play at this "for hours." Janna thinks that Finley believes she is invisible when her face or her head is covered.

Janna's interpretation is slightly different from what we have heard from other informants, who describe the child as thinking the adult has disappeared when out of sight. Apparently, as the child starts to assume the role of director, so also does the adult begin to reimagine the nature of the illusion. Rather than the adult disappearing, now that the child can cover herself, she is also able to conceive of herself as disappearing. Quite flexible in their interpretations, these adult illusionists!

During this same age period of five to eight months, shortly after establishment of the mutual form of Peekaboo, the second of the two fluid stages mentioned above emerges when the adult conceals and reveals an object like a toy. Alexander, one of Ratner and Bruner's subjects, had played "direct" Peekaboo two months before starting to play with a clown toy hidden by his mother (1978, 393). The hiding is accompanied by the alert and release calls familiar to the child from Peekaboo proper. Here the mutual gaze is supplemented by the triangulated gaze of joint attention as the parent directs the child's attention toward the toy with vocalizations and movements of the toy, shaking it or moving it closer to the child's face. Interestingly, in this frame the adult and child might make eye contact while the object is out of sight, then the adult will see the child turn her gaze to seek the object at the place the child anticipates it will reappear. It is evident that the immediate goal of play is not to establish the mutual gaze, as in Peekaboo proper, but to focus on mentally contacting the hidden object.[18] Sometimes the infant will attempt to move toward the toy or reach for it, ratifying the adult's sense that the toy has the infant's attention, and thus the performance position for the start of another round is established. Several researchers have noted that once a round or two has launched a game, the reappearance of the object in different locations seems to violate the younger child's expectations, not eliciting the same joyful response (e.g., Charlesworth 1966; Parrott and Gleitman 1989). The folk illusion morphologies for the two versions of Peekaboo—with and without a toy—are very similar, varying only in the object of attention and other attendant positional elements such variation entails.

6a. Agent establishes performance space and time by drawing child's attention to the toy by shaking or moving it, with joint gaze and/or alert call
6b. Agent hides the toy
6c. Priming period
6d. Agent reveals the toy / release call
6e. Patient sees the toy
6f. Agent and patient respond to the reappearance

Later, beginning around nine and continuing until fifteen months, the form of play changes again as the child begins to take over the role of play initiator. Caretakers may facilitate this by, for example, placing a towel or cloth over the child's face in such a way that the child can pull the cloth off herself and reappear. The child may opportunistically initiate Peekaboo at a moment when eye contact with the adult is inadvertently broken, maybe when something covers the baby's face, much like in the earlier form of the game when the adult would seize the moment of a chance interposition between the two. In folk illusion terms, the child is growing into the role of the director, increasingly able to control mutual gaze by actively finding or creating opportunities to break the gaze, such as averting her face from the adult, covering her own or the adult's face, or even moving into a hiding position. The adult, sensitive to the child's attempts to key a Peekaboo performance, adapts to the child's play.[19]

Where's Zoa?

Claiborne Rice and K. Brandon Barker

We were most fortunate that Brandon and his wife, Ellen, had a beautiful little girl, Zoa, in the fall of 2014. Like many contemporary families, Brandon and Ellen have amassed an impressive catalog of candid home-video footage, captured on cellular phones. Once we became interested in Peekaboo, we scoured videos of Zoa and gathered instances depicting Zoa playing Peekaboo with her parents, other relatives, and friends between the ages of ten months and thirty-six months.

Several videos of Zoa beautifully demonstrate the adults' adaptiveness to the child's attempts to key performances of Peekaboo. In one, from November 2015—Zoa is almost fourteen months—the video opens with Zoa's having covered nearly her whole body with a large, white towel. She is performing before several adults, who all share in the alert calls of "Where's Zoa?" As Zoa takes a step forward, the towel slides off. With her face and body now exposed, the adults say "There she is" and clap. Zoa looks up and around at the adults, smiles widely, then claps also, as if to say, "Hey, this is fun!" As she bends to grab the towel again, the alert call is already coming—"Where's Zoa?" She manages to get her arms up enough so the towel covers her eyes, but as she steps forward she treads on the towel so it immediately falls off again. Her face is covered for less than a second, but the release calls echo nonetheless—"there she is!" She walks around and claps, and the adults clap and say, "Yay!" Now when she circles back around to the towel on the floor, no alert calls are heard until she bends down and gets the towel back up in front of her face. At just that moment we hear some cries of "Where's Zoa?" She manages to get two steps before dropping the towel, this time almost two full seconds. The ease with which she drops the towel and turns back to the crowd, clapping and smiling, to acknowledge the release calls suggests that she has dropped the towel this time in anticipation of the calls and clapping. By holding the alert and release calls until the appropriate moments, the gathered adults have allowed Zoa to become the director of the activity. The question of her hiding or the adults having disappeared has become unimportant as the form of play gets carried out as a performance, the attendant delights arising from the carrying out of the activities and the approbation of the family more than from the detection of an illusion of absence.

Just a few days later, Zoa plays Peekaboo within a curtain, a covering the child can manipulate more easily than the large towel, and thus the timing of the game seems more carefully controlled by the child. The thirty-seven-second video opens with Zoa behind a curtain calling out, "Yes!" Her mother is saying, "Where is she? I can't see her. Is she under the bed? No." At this point, four seconds in, Zoa swings the curtain open to reveal herself smiling. "Ah. There she is! Hee hee. I found her," says Ellen. Ellen draws out the sentence, "There she is," beginning at an unusually high pitch, then falling dramatically on "is"—immediately recognizable as the Peekaboo release call. Right away, Zoa says, "Again!" and swings the curtain to hide herself. Ellen repeats as confirmation, "Again?" The game is afoot. "Where's Zoa? Is she under Daddy? No." Zoa has been hiding for five seconds when she interrupts Ellen's "No" by swinging the curtain around, seeking eye contact. Ellen gives a sharp, audible intake of breath, followed by a full-second dramatic pause, then says, "There you are!" with the same drawn out, Peekaboo release-call contour. "Who's that?" says Zoa, and waves. "Oh, hi," says Ellen, showing that she understands Zoa's gesture as a hello wave. "Are you gonna hide again?" "Yes," says Zoa. Then Ellen says "again" in the same tone she used at the outset of the second round. The word uttered in this form seems to key the performance on Ellen's part. Round three begins with Zoa pulling the curtain around to hide herself and saying, "Yes!" "Where's Zoa, where's Zoa?" Ellen responds. Zoa yells something, then Ellen says, "Daddy, where is she?" The curtain swings open—there is Zoa seeking eye contact, having hidden seven seconds this time. Ellen gives the sharp intake of breath and the release call. (See Video 1.)

The video ends here, but one gets the idea that it might have gone on much longer. Isolated behind the curtain, Zoa gives the impression of someone both hiding and aware of her isolated status. Like the adults a few days before, Ellen takes her cues from Zoa's actions. Zoa's firm control of the curtain as opposed to the towel, however, seems to put Zoa more effectively in the director's chair, and thus the sense is more palpable that she is playing with a feeling of either her own seclusion from her parents or her delight with their reappearance. Considered in the context of other folk illusions, these instances of Peekaboo reinforce that body acquisition is not simply or only the acquisition of the particular skills of controlling the curtain's movement, or of using words to key the performance, or even of registering the performance keys provided by her mother. The element of timing demonstrates all of these particular skills deployed in an unselfconscious way that amounts to moment-to-moment living, to a way of being in the world that responds to the physical but also psychosocial and even spiritual features of her world. The love and safety felt in both performances are as much a feature of the landscape as the bedroom curtain itself. If skills and abilities are sedimented over time, to use Merleau-Ponty's term, then the whole of the performances, in all of their particularities and intangibles, informs the child's embodied being.

Like all other folk illusions, Peekaboo has a priming period, an important rhythmic element that emerges from the play of all participants, responding primarily to the actor's body's adapting to the performance requirements as play proceeds, but also sensitive to the various contingencies, like the number of participants or the nature of the play space. When we watch kids of all ages play Peekaboo, the priming period is unmistakable. Jerome Bruner, in his two landmark studies (Bruner and Sherwood 1976; Ratner and Bruner 1978), described the timing of the constituent parts of Peekaboo play as its most intriguing feature.[20] In 1993, Fernald and O'Neill lamented that "no studies to date have described in detail how mothers time their reappearance in relation to alert calls and to the duration of hiding, nor how this temporal variability influences infants' persistence and pleasure in playing the game" (278). A close study of the timing in Peekaboo has not yet been conducted, but comparison with other folk illusions is instructive. In early forms, the adult agent controls the amount of time that the gaze remains broken. Adults seem sensitive to the child's level of discomfort with having the face covered. This is not based merely on visual feedback, as an adult with her face covered or her hands over her eyes cannot readily see the child (though peeking is usually allowed!). Discomfort or agitation can be detected auditorily or even tactilely if the adult is holding or touching the baby.

The notion that many adults have that the baby thinks the adult has disappeared also informs timing. Intuitively adults know roughly how long it takes to scan a room or an area for an item and deduce its absence. If the adult imagines the child is carrying out such a scan, then the adult can wait that necessary moment before the reveal. As children become directors in Peekaboo play, the priming period becomes even more apparent. The adult will wait for the child to lower the covering implement or turn her head to establish the mutual gaze before giving the release call, "There she is!" Sometimes the child simply has trouble coordinating her arms to accomplish the uncovering, but any difficulty is swept up by the adult seamlessly into the game.

Pleasures of Perception in the Aggregate

Success is a formal quality of folk illusions, but it is attested to by the emotional qualities of pleasure and fun. Folk illusions are able to be traditionalized because the form of the activities must take particular shapes in order to succeed at prompting illusions. The body is a space in which certain moves effect certain responses and in which only a subset of those responses are unexpected, sustainable, and present a stable enough morphology that they can be replicated by other bodies. It is worth examining each form and asking what the pleasure of the form is and what counts as a successful iteration of a performance of it. Most folk illusions reproduce physical effects that are unfamiliar and unexpected to some degree by naive actors, yet, like a good poem, are remarkable enough to be enjoyed even when the effects are anticipated.

The optico-geometrical illusions illustrate this aesthetic dynamic well. One can be reasonably certain that if the target figure is drawn within certain parameters, a viewer will experience the discordant perception. Likewise, the correct performance of a Kohnstamm illusion will result in muscular activation and the concomitant proprioceptive sensations.[21] Also like the optico-geometric illusions, there is a range of strength of effect, such that some people more easily feel or more readily recognize certain somatic responses. In Aristotle's illusion, for example, the inner Sherlock's assumption that the outsides of two fingers must be sensing two different objects, or two different places on an object, is so strong that for those who experience the illusion, visual confirmation that one is touching one's own nose is frequently not enough to change the sensation of touching two noses. We have witnessed on many occasions people being spooked or freaked out by the unexpected sensations arising from performing a folk illusion, while others—adults and children—express a milder surprise or curiosity. Indeed, we have spoken with people who had tried a variety of folk illusions before but found them utterly ineffective. Individual variation in responses to illusions is a regular feature of the genre.[22]

Another set of forms, the coordination illusions, evokes a different kind of surprise, associated perhaps with a gap or lacuna in one's body map, such that certain nominally unrelated motions cannot easily be performed together.[23] The most familiar of these forms is Pat and Rub (E5). Rubbing one's belly in a circle and patting one's head are individually easy actions to accomplish, but when attempted simultaneously, the movements create a conflict that renders the independent motions difficult.[24] The coordination illusions challenge any easy assumptions about how success and enjoyment work in folk illusions. It stands to reason that the goal of Pat and Rub is to succeed at patting one's head and rubbing one's tummy at the same time—indeed, that is the stated goal when directors introduce the form to others. But surprisingly it is the failure of some group members to achieve success that makes playing it fun. In our observations of young children between ages four and eight, individuals declare success at the operation despite what we as observers see as the lack of successfully coordinated motions. We understand the children's claims as reflecting their awareness of the social context—that is, the understanding the kids have of the goal of the activity in the group. Older children display the smiles and laughs of surprise and enjoyment when they and their friends cannot execute the motions, and sometimes they heighten the intensity of their focus as they work against their own body to coordinate the motions of their two hands. There is a certain mystery of the body lying at the heart of one's strange incoordination that eventual mastery of the activity does not fully dispel.

Based on our observations and collected remembrances, we know that often younger kids first attempt a coordination illusion when prompted by older siblings or teachers. This introduction may suffice to activate the form, but horizontal transmission cannot be achieved until the children have matured beyond a simple desire to imitate or execute the moves correctly and can begin to appreciate the mysterious resistance of the body more reflexively. Thus, success in the performance of coordination illusions is paradoxical—it truly occurs only when an actor both fails to execute the directed motions and also realizes that this failure is mysterious or unpredicted.

In Finger Twirl (E7), the actor attempts simply to twirl her two index fingers (or whole fists, in one variant) in opposite directions while they are pointed at one another in front of the actor. (Of course, the term *opposite* is relative! In one sense, the fingers are to be twirled in the same direction. For example, each finger should be moving in a clockwise direction when viewed down the longitudinal axis of the extended index finger.) Try it! Most adults will find this achievable only with considerable difficulty. One feels as though one is exerting a great deal of mental force to overcome the body's enigmatic resistance. It is this force that becomes the background of attention, even as one concentrates on the shape of each finger's motion, perhaps triumphing eventually in execution. Similarly, cataloged as a variant of Twirling Fingers is Impossible Signature (E8), in which a seated actor is directed to swing one leg circularly in the air, rotating from the knee. After a good rhythm is established, the actor is directed to write his signature in the air with the index finger of his dominant hand. These coordinated actions prove surprisingly impossible to execute simultaneously. The leg's circular trajectory usually distorts on the first letter of the signature, often unintentionally following the signature in shape as well. Strong concentration is needed for adults to accomplish even a single letter while maintaining the circular motion of the leg (unless their name is OO!). Everyone we have observed performing this illusion for the first time is astonished at its difficulty, and most try several times to accomplish it.

One pleasure of this form for the director resembles that of the riddler able to stump his interlocutor, but in this case the actor cannot eventually savor the clever answer. Instead, the body itself is posed as a riddle that its inhabitant cannot solve. The point of knowledge about the body held by the director does not extend to the physiological mastery of the professional doctor; it lies simply in knowing this hidden conflict of the nerves.

Children and youths naive to particular folk illusions have an odd gap in their theory of the body. They cannot possibly know what another person is feeling if they do not know that such an experience even exists. It is like asking someone how Johnny felt when he was splaked by the kazoo. What is *splaking*? What is a kazoo? How do kazoos splake? Anyone familiar with fire can imagine what Johnny felt when he stood too close to the campfire that time and his boot caught on fire and started melting and they had to take him to the emergency room. But even lifting our own arms (which most people will have done thousands of times by the age of ten) cannot prepare us to know what it would be like for our arms to lift themselves like zombie arms, acting all on their own. Our bodies aren't so unruly! The source of the illusion in coordination play is exactly this: kids (and most adults) firmly do not believe that they cannot do Pat and Rub. They are certain they can do it. We saw a whole group of kids under the age of ten insist that they could do it. Then they started patting and rubbing, and, of course, none could do it in the short time we had to observe. The kids who were aware that they did not successfully pat and rub learned a secret about their own and other bodies. They were not in pain, and they did not tax their bodies painfully to some physiological limit. This is the *eu*-appearance of the body, the thematizing of the body as an object of mystery and fun. Again, this may account for the unusual joy we encounter every time we mention these illusions to kids or adults. From our perspective, having talked to hundreds of people of all ages, the mention of a folk illusion is a prompt to play, to memories of joy and fun. Johan Huizinga observed that it is "this fun-element that characterizes the essence of play. Here we have to do with an absolutely primary category of life, familiar to everybody at a glance right down to the animal level" ([1944] 1949, 3).

It is the nature of play to stand outside of our serious pursuits, to be downplayed as trivial. But maybe the triviality barrier exists to preserve the realm of play for us as a place of lightness and freedom. If so, folk illusions can send us there for a moment, as our body itself remembers and reappears in its oddness as that which always just escapes the grasp of the known.

Notes

1. The essentially ambiguous, antithetical quality of play is a major theme running throughout Sutton-Smith's work. He most clearly articulated the idea, however, in *The Ambiguity of Play* ([1997] 2001).

2. Bauman presents the term *adultocentrism* on analogy from ethnocentrism in the contexts of the difficulties of education in his article "Ethnography of Children's Folklore" (1982, 173–74).

3. The Harvest Music Festival was held annually from 2006 to 2014. Clai Rice attended this festival three years running, and this observation occurred during the weekend of October 17–19.

4. Briefly, there exists a zone, called Panum's area, extending from a little closer than your focal point to a little farther than your focal point, in which the images on each retina appear fused into a solid object. Regarding objects outside of Panum's area you experience

diplopia, or a double image. Objects farther away than Panum's area appear in the zone of uncrossed disparity because you would have to uncross or diverge your eyes to achieve fusion; objects closer than Panum's area are in the zone of crossed disparity because clear focus would require your eyes to converge. We should point out that there are likewise two different forms of the Floating Finger illusion. Holding your fingers in front of your eyes and crossing your eyes puts your fingers in the zone of uncrossed disparity, but looking past your fingers puts them in the zone of crossed disparity. The two variants are functionally the same; as far as we have been able to tell, there is no folk awareness of this difference.

5. Daw lays out the developmental course of the various properties contributing to normal human vision (2005, 188–89). Some properties, such as the ability to detect a small displacement between two line segments (Vernier acuity), do not reach adult levels until the age of ten years, while others, like the time it takes to adjust to very dim light, are already the same in infants and adults (41). Convergence (movements of eyes to objects at different distances) and accommodation (changes in the shape of the lens to focus images on the retina at different distances), both properties important to the experience of Floating Finger, are largely adult by age four and a half, but their duration is not fully adult until age eight (Daw 2005, 187; Yang and Kapoula 2004, 223). Suppression, the mechanism by which the image in one eye is suppressed in favor of the image in the other eye, is the visual system's method of avoiding diplopia. Floating Finger manipulates the nature of suppression. One eye's image of each finger is suppressed until the tips are superimposed, triggering the unsuppression of the overlapping area. Daw states that the "period for development for suppression is unknown" (2005, 187). Evidence from this illusion suggests that researchers should expect it to reach adult levels at around age eight.

6. Experimental studies of development have focused, at times, on illusion perception. Piaget's *Mechanisms of Perception* (1969) features empirical investigation of optico-geometric illusions across age groups. There, he makes a parallel argument to ours, which is that sensitivity to some types of illusions, such as the Oppel-Kundt illusion, increases with age and development. See, for example, "Perceptual Activities and Secondary Illusions" ([1969] 2013, 137–45).

7. We met a five-year-old child in Bloomington, Indiana, who could in fact kiss his elbow, so there was no illusion for him!

8. For a brief explanation of Chomsky's view on Universal Grammar and development, see *The Science of Language* (Chomsky 2012, 153–56).

9. See Schoneberger 2010 for a summary of the *interactionist* view of acquisition, especially for language. Interactionism avoids reductionist arguments from heredity while aiming to provide accounts of the acquisition of specific skills that acknowledge the contributions of both genetic and environmental factors.

10. Dolezal (2015, 25–30) mentions later phenomenologists like Drew Leder and Shaun Gallagher but focuses her discussion of the "invisible body" on how important it is for Sartre, who, of course, sees self-consciousness as resulting from one's body being seen by the Other.

11. Dreyfus points out that Merleau-Ponty uses the terms *skill* and *habit* interchangeably, so that "the ability to perceive is like an already acquired bodily skill" (1999, par. 5).

12. See for examples, Enfield, Majid, and van Staden (2006, 141) and Witkowski and Brown (1985).

13. Gibbs (2017, 205) summarizes the extensive cognitive scientific research on how using metaphorical action phrases like *grasp the concept* invokes sensory motor regions of the brain also used to plan and categorize the physical motions involving the relevant part of the body.

14. Washburn (1929) studied mothers and children playing Peekaboo as part of a larger study of infant smiling and laughter. Goldsmith and Rothbart (1999) include Peekaboo routines as part of a large battery of tasks to assess infant social positivity both before and after the children can move around on their own. This assessment has been widely used and adapted in child development studies. Other researchers have developed Peekaboo tasks as research paradigms for examining adult-child interaction, emotion expression, or social development; see, for example, Montague and Walker-Andrews (2001).

15. Fernald and O'Neill refer to the two basic forms as forming "fluid stages" (1993, 269). In the psychology literature, the mutual gaze form has been identified as appearing initially, before forms involving hiding another object. The two forms then develop along a similar trajectory and are understood by researchers as two forms of the same game; see especially Ratner and Bruner (1978, 396), where a game of hiding a toy is understood to give way seamlessly to a form with the mother hiding herself behind a chair, then the child hiding himself behind the same chair.

16. Ratner and Bruner defined the terms used by subsequent researchers: "We defined a ROUND as one complete cycle of . . . disappearance and subsequent reappearance, and a GAME to consist of any uninterrupted sequence of rounds" (1978, 393).

17. According to Ratner and Bruner, adults frequently state that they hide their own faces rather than cover the child's due to their sensitivity to the child's becoming afraid (1978, 284).

18. Langacker, anticipating the important role of joint attention in current theories of language development, defines the psychological notion of mental contact: "a person makes mental contact with [item] t_i when, in his current psychological state, t_i is singled out for individual conscious awareness. When both S[peaker] and H[earer] make mental contact with t_i, full coordination of reference is achieved" (1991, 91). In Peekaboo, the adult's mature understanding of mental contact gets projected onto the infant—the adult assumes that both participants have made mental contact with the same toy, and, through this triangulation, infers that their respective current mental-space constructions correspond at least to this limited extent.

19. See Fernald and O'Neill's analysis of the movement from passive to active participation in Peekaboo interactions (1993, 271–75).

20. These two studies are widely cited by scholars of Peekaboo in psychology.

21. Using our definition of *illusion*, by which many of our perceptions are nonveridical in a strict sense, asking how many people do not experience illusions would be nonsensical. Everyone experiences some illusions as part of normal perception. It is reasonable, however, to ask how many people might or might not experience any particular illusion, given certain parameters. This work has been done for many of the optico-geometric illusions, especially in the context of testing the Carpentered World Theory, mentioned in chap. 1; see Deręgowski (2013) for an excellent summary. Very few folk illusions have been investigated this way. De Havas, Gomi, and Haggard (2017) report that Floating Arms is experienced by 75 percent of subjects when prescreening is absent.

22. While performing folk illusions in Bloomington, Indiana, in the fall of 2016, a nine-year-old participant reacted so aversively to Dead Man's Hand that he no longer desired to participate in the activities. Hannah Ritorto, during our interview, literally jumped from her seat when Concentration was mentioned. She hates the feeling of the chills so much that the mere memory of them provoked a visceral response. She remembered playing Concentration with her friends but she never allowed them to go past the crack-an-egg stage on her. A student of Clai's, around thirty years old, stepped from a doorway during her first

performance of Floating Arms and collapsed on the floor after feeling her arms float up. She refused to stand up for several minutes because she was "afraid [her] arms will move by themselves."

23. Fauconnier and Turner ([2002] 2008) state that image schemas are nonpropositional, but that one can run inferences over them; by this notion, one can run inferences over one's body schema. Can you put your foot in your mouth, can you twist your arm twice around your body, can you turn your head all the way around like an owl—all these questions should be answered in the negative even if the actions have never literally been tried; trials can perhaps be simulated, but that might be what running inferences over a schema means. Anyway, these lacuna would appear to exist because of a shortcoming of the body schema, or the inference running mechanism. This question is important in the context of development, because at some point these schemas or the inferencing process do become more accurate.

24. Pat and Rub is frequently mentioned in technical discussions of interference, a highly researched phenomenon. *Bimanual interference* is the term used for the way that certain noncomplementary motions cannot easily be performed by both hands simultaneously. Research has disclosed some interesting characteristics of this effect, so, for example, goals given factually—for example, as a visual presentation of a target—do not cause interference but goals given symbolically do. For a recent study of bimanual interference across hand actions in children ages four to eleven, see Otte and Mier (2006).

APPENDIX
Catalog of Folk Illusions

A. Haptic Illusions

A1. Crossed Fingers

Old and well known, this illusion is often referred to as Aristotle's illusion. The middle and index fingers of one hand are twisted. Then the actor closes her eyes and touches her nose. Most people report the sensation of having two noses.

* *See our discussion of Aristotle's illusion in chap. 4; Benedetti (1985).*

A1A. CROSSED FINGERS, SPLIT TONGUE

Assuming the same crossed-finger position of A1, the actor touches her tongue, which then "feels like your tongue is split in half."

* *Remembrance gathered from a folklore student, spring 2014, played as nine-year-old, Westfield, New Jersey.*

A1B. DOUBLE CROSSED FINGERS

We observed a variant of Crossed Fingers in which the actor crosses the middle and index fingers of both hands and touches those fingers tip-to-tip without looking at her hands. She then twists her hands back and forth so that it is difficult to *feel* which tips of which fingers are touching which tips of which fingers.

* *Observation of Monica Hesse, summer 2009, twelve-year-old, Lafayette, Louisiana; remembrance gathered from a folklore student, spring 2015, played as seven- or eight-year-old, Columbus, Indiana.*

A2. Where Am I Touching You?

(See Video 2.) The patient stretches out one of her arms, palm up. The agent pinches the patient's forearm near the elbow crease, then tells the patient to close her eyes. With the patient's eyes remaining closed, the agent slowly and gently runs the tip of her finger in a circuitous path from the patient's wrist up the arm toward the previously pinched spot. The patient is instructed to state when the agent has arrived at the pinched spot. Invariably the patient calls arrival well before the agent has actually arrived at the spot. When the patient opens her eyes, she is mystified as to why she felt that the spot had been reached.

* *Observation of E. Hesse, age twenty-three, and Emma Tomingas, age twenty-five, spring 2009, Lafayette, Louisiana.*

A2A. WHERE AM I TOUCHING YOU? (NO PINCH)

The agent taps alternatingly the index finger and middle finger as he moves from the wrist to the crease of the elbow. The agent *does not* pinch the crease of the elbow before beginning. The patient's actions are the same as A2.

* *Remembrance gathered from a folklore student, spring 2016, played as twelve-year-old, Cincinnati, Ohio; see also Brugger and Meier (2015), who report a Swedish version that leaves out the pinch in their experimental study, "A New Illusion at Your Elbow."*

A2B. FIRE TRUCK

Taking the performance position described in A2, the agent states that his fingers are a fire truck and that the patient is standing in the street at the crook of her elbow. She should yell "Red light!" when the fire truck gets close to her. He runs his fingers slowly along the forearm. When she yells, the agent pops her on the forehead and says, "Fire trucks don't stop at red lights!"

* *Remembrance gathered from a folklore student, spring 2016, played as six-year-old, Hong Kong.*

A3. The Chills

(See Video 3.) The agent stands behind the patient and recites the lines shown under the following subvariants while performing the correlative actions (in brackets) upon the patient.

* *Observation of four twelve-year-old girls, summer 2009, Lafayette, Louisiana; observations at St. Cecilia Middle School, spring 2011, Broussard, Louisiana. Mary and Herbert Knapp record a variety of similar lyrics for an unnamed "ceremony that is also a trick." (1976, 250).*

A3A. CONCENTRATION

1. Concentration, concentration.
2. Crack an egg on your head. [*Agent places a fist on top of the patient's head and slaps fist with the other hand.*]
3. Let the yolk run down, let the yolk run down, let the yolk run down. [*Agent runs fingers down patient's head, neck, and back.*]
4. Concentration, concentration.
5. Stab a knife in your back. [*Agent pokes patient between the shoulder blades.*]
6. Let the blood run down, let the blood run down, let the blood run down. [*Agent runs fingers down patient's back.*]
7. Concentration, concentration.

A3B. EIGHTY DAYS AROUND THE WORLD

1. Eighty days around the world. [*Agent draws a circle on the patient's back with the tip of her index finger.*]

2. X marks the spot. [*Agent draws an X on the patient's back with the tip of her index finger.*]
3. Comma, comma, comma. [*Agent draws three short vertical lines across patient's shoulder blades.*]
4. Question mark. [*Agent draws a question mark in the same fashion.*]
5. Spiders going up your back. [*Agent crawls both hands from patient's lower to upper back.*]
6. Spiders going down. [*Agent crawls both hands from patient's upper to lower back.*]
7. Spiders going up your back. [*Agent crawls both hands from patient's lower to upper back.*]
8. Spiders going down. [*Agent crawls both hands from patient's upper to lower back.*]
9. Tight squeeze. [*Agent pinches the trapezoid area of patient's shoulders.*]
10. Cool breeze. [*Agent blows on the back of patient's neck.*]

A3C. FORK IN YOUR BACK

1. Fork in your back. [*Agent pokes three or four fingers in patient's mid-back.*]
2. Let the blood run down. [*Agent strokes all fingers down patient's back.*]
3. Let the chills run up. [*Agent strokes all fingers up patient's back.*]
4. Egg on your head. [*Agent places fist on top of patient's head and slaps fist with the other hand.*]
5. Let the yolk run down. [*Agent strokes all fingers down patient's head, neck, and shoulders.*]
6. Let the chills run up. [*Agent strokes all fingers up patient's back, shoulders, neck, and head.*]

A3D. PEOPLE ARE DYING

People are dying, children are crying.
Concentrate.
People are dying, children are crying.
Concentrate.
Crack an egg on your head.
Let the yolk run down, let the yolk run down, let the yolk run down.
Crack an egg on your head.
Let the yolk run down, let the yolk run down, let the yolk run down.
(Chorus again)
Crush an orange on your head.
Let the juice run down, let the juice run down, let the juice run down.
Crush an orange on your head.
Let the juice run down, let the juice run down, let the juice run down.
(Chorus again)
Stab a knife in your head.
Let the blood run down, let the blood run down, let the blood run down.
Stab a knife in your head.
Let the blood run down, let the blood run down, let the blood run down.
(Chorus again)

Fig. A.1. Calvin (age six) and Jacob (age nine) perform Dead Man's Hand.

A4. Flayed Hand

The agent tightly holds one of the patient's wrists with one hand. With the other hand, the agent aggressively rubs the open-handed palm of the same wrist. After rubbing for a priming period of approximately twenty to thirty seconds, the agent peels her rubbing hand free from the patient's palm. The patient feels as though the skin of her palm is being peeled off.

* *Remembrance gathered from a folklore student, spring 2015, played as seven- to thirteen-year-old, Cincinnati, Ohio.*

A5. Dead Man's Hand

(See Video 4.) The experiencers place their index fingers together from tip to base (or hands together, palm to palm). Then, one actor rubs the "tops" of the two fingers vertically with his free index finger and thumb. Each actor's fingers feel as though they are a part of the other's body, and they feel coarse and numb—or "dead." Once the performance position is assumed, participants usually take turns rubbing the fingers until the effect is achieved.

* *Remembrance gathered from a folklore student, spring 2015, played as six- to seven-year-old with her father, Indianapolis, Indiana; remembrance gathered from a folklore student, fall 2013, played as seven-year-old, Austin, Texas; another folklore student, spring 2014, played as sixteen-year-old, Beech Grove, Indiana; see also Boulware (1951), Arnold (1952), and Dieguez et al. (2009).*

Fig. A.2. Eight-year-old Aoife performs Touching Invisible Glass.

A6. Touching Invisible Glass

(See Video 5.) The experiencer places his hands in front of his chest in a praying position, but he only touches the fingertips together. Then, the experiencer moves the palms of the hands closer together and farther apart in succession for a priming period of about thirty seconds. After the priming period, the experiencer feels as though a pane of glass is between the touching fingertips of each hand.

* *Remembrance gathered from a folklore student, spring 2014, played as ten- to twelve-year-old, Greenville, South Carolina; remembrance gathered from a folklore student, spring 2014, played as nine-year-old, Wayne, New Jersey; remembrance gathered from a folklore student, spring 2014, played as seven- to twelve-year-old, Basking Ridge, New Jersey.*

A7. Furry Air

(See Video 6.) The director instructs the actor to position her hands in front of her body so that the fingertips of each hand are pointed at the correlative fingertips of the other hand. The fingers should also be spread apart as if each hand is holding an invisible softball-sized ball. The actor is then instructed to rapidly move her arms so that her hands oscillate back and forth while keeping the fingertips of each hand pointed at one another. The actor's hands will feel as though they are moving through furry air, as if "petting a dog."

* *Remembrance gathered from Daniel Povinelli, age forty-seven, fall 2010; remembrance gathered from a folklore student, fall 2013, played as ten-year-old, north Indiana.*

A7A. MAGIC BALL

The actor assumes the same performance position and performs the same actions as A7, but instead of reporting the sensation of furry air, the actor reports the sensation of an invisible "magic" ball, or softball.

* Remembrance gathered from a folklore student, spring 2014, played as ten- to thirteen-year-old, Long Island, New York; remembrance gathered from a folklore student, fall 2013, played as eight- to twelve-year-old, Iowa City, Iowa.

A8. Electricity at the Fingertips

(See Video 7.) The actor forms her hands into loose fists so that she can rub the tops of her fingernails against each other, back and forth for a priming of twenty to thirty seconds. After rubbing, the actor allows her fingertips to hover close together, feeling as though there is electricity between them.

* Observation at St. Cecilia Middle School, spring 2011, Broussard, Louisiana.

A9. Electricity in the Palm

The director tightly squeezes one of the actor's wrists. The actor is then instructed to forcefully open and close his wrist ten to twenty times. The director then releases the actor's wrist, and the actor perceives a feeling of electricity in the palm.

* Remembrance gathered from a folklore student, fall 2013, played as eight- to ten-year-old, South Korea; remembrance gathered from a folklore student, spring 2015, played as six- to seven-year-old, Seoul, South Korea.

A10. Shockers in the Palm

The agent vigorously rubs a patient's palm and fingers for a priming period of about thirty seconds. Then, the agent pinches the center of the rubbed palm. The patient perceives "the shockers" in the palm and hand.

* Remembrance gathered from a folklore student, fall 2013, played in the third grade, Louisville, Kentucky.

A11. String Pull

(See Video 8.) The actor presses his right thumb into the center of his right palm. He then clenches his fist tightly around this inserted thumb. The director then slaps the actor's knuckles on his right hand one hundred times. After this, the director instructs the actor to open his hand and pulls an imaginary string through his palm. The actors feels as though a string *is* being pulled through his palm.

* Observation at St. Cecilia Middle School, spring 2011, Broussard, Louisiana.

A11A. STRING PULL (FRICTION)

The agent forcefully smooths out the creases of the patient's palm until it is red. The agent then lightly touches all of the creases of the agent's palm. After that, the agent lightly pinches the middle of a crease, pulling as little skin as possible. Then, the agent pull her fingers away, pretending to be pulling a string out.

* Remembrance gathered from a folklore student, spring 2014, played in elementary school, Palatine, Illinois.

A12. Who's Touching You?

A very well-known, highly attested folk illusion. The agent, who is positioned behind and on one side of the patient (outside of the patient's field of vision), taps the patient's shoulder on the opposite side. That patient automatically looks behind herself to the side of the touched shoulder. It appears that she has been touched by no one.

* *See "Is Sammich Touching You?" in chap. 4.*

A13. Force between Your Fingers

The actor places her index fingertips as close together as possible. After an appropriate priming period, the actor perceives a force, like "the opposite ends of magnets" between the two fingertips.

* *Remembrance gathered from a folklore student, fall 2013, played in middle school, Goshen, Indiana.*

A14. What's Tickling You?

The director instructs the actor to close his eyes. Next, the director is asked where he feels a tickling sensation. To the actor's surprise, she does begin to feel a tickling sensation somewhere on his body. The director then instructs the actor to say what is causing the tickling sensation. The actor usually describes some "insect, spider, or creepy-crawler." Unbeknownst to the actor, a third participant has been very lightly blowing on the area that feels tickled.

* *Remembrance gathered from a folklore student, spring 2015, played in middle school, Binan, Philippines.*

A15. Knock the Quarter off Your Head

The director forcefully presses a quarter (or some other coin) on the actor's forehead and slyly removes the coin, acting as though the coin is still stuck to the actor's forehead. The director then instructs the actor to try and hit the coin off of their forehead without touching it. Because of the haptic aftereffects of having the coin pressed against his head, the actor believes as though the coin is still there and "ends up hitting the back of his own head when the coin isn't even there."

* *Remembrance gathered from a folklore student, spring 2016, played as ten-year-old at school (he had learned the trick from his father), Lafayette, Indiana. See also the children's book* Magic and Card Tricks *by Jon Tremaine, where the illusion is referred to as* brow beating *(2002, 38–39).*

A16. The Almost Nose Touch

The agent hovers her pointing index finger just in front of the bridge of the patient's nose. The patient cannot avoid feeling a "weird" and "tickling" sensation.

* *Remembrance gathered from a folklore student, fall 2013, played between fifth and seventh grades, Chicago, Illinois; remembrance gathered from a folklore student, spring 2016, played as ten-year-old, Mumbai, India.*

B. Visual Illusions

B1. Rubber Pencil

A pencil is lightly held between the thumb and index fingers. The pencil is held in front of the actor and "wobbled" so that the pencil appears to bend back and forth.

* Observations at St. Cecilia Middle School, spring 2011, Broussard, Louisiana; remembrance gathered from a folklore student, spring 2015, played as ten- or eleven-year-old, Breckenridge, Colorado; see also Pomerantz (1983).

B2. Detachable Thumb

The agent's thumb of the left hand is bent at the top knuckle, and the index finger of the same hand is folded over the bent knuckle. The thumb of the right hand is similarly bent, and the knuckles of the two thumbs are brought together under the right index finger to provide the illusion that the two thumbs are actually only one (the joining point is hidden by the right index finger). The two hands are then pulled apart and rejoined; this gives the illusion that the top of the right thumb is being pulled off.

* Remembrance gathered from a folklore student, spring 2014, played as seven- to fourteen-year-old, Atlanta, Georgia; remembrance gathered from a folklore student, spring 2014, played as ten-year-old with uncle, Bangkok, Thailand; remembrance gathered from a folklore student, spring 2014, Dubai, United Arab Emirates; see also Welsch (1966, 176).

B3. What Did I Say?

The agent mouths certain words that the patient interprets as something more recognizable.

* See "What Did I Say? I Love You, Elephant Shoe, and Olive Juice," chap. 3.

B3A. ELEPHANT SHOE

The agent, after finding an appropriately intimate moment during which she is making eye contact with the patient, mouths the words *elephant shoe*. For the patient, the agent appears to be mouthing *I love you.*

* Remembrance gathered from a folklore student, fall 2013, played as nine- and ten-year old, Chicago, Illinois; remembrance gathered from a folklore student, fall 2013, played as eight-year-old, Trafalgar, Indiana.

B3B. OLIVE JUICE

The agent, after finding an appropriately intimate moment during which she is making eye contact with the patient, mouths the words *olive juice*. For the patient, the agent appears to be mouthing *I love you.*

* Remembrance gathered from Tina Mitchell, age thirty-two, summer 2011; remembrance gathered from a folklore student, spring 2015, played in elementary school, Minneapolis, Minnesota.

B3C. VACUUM

The agent, after finding an appropriately intimate moment during which she is making eye contact with the patient, mouths the word *vacuum*. For the patient, the agent appears to be mouthing *fuck you*.

* *Remembrance gathered from a folklore student, fall 2013, played as twelve- or thirteen-year-old, Provo, Utah.*

B4. Roped Off

The agents stand on either side of a pathway (e.g., a sidewalk, hallway, or road). The agents kneel and position their hands as though they are holding a rope or line across the pathway. The patient, whether walking, riding, or driving through, reacts to the illusory blockage stretched by the agents. Patients often step over the invisible rope or slow down their automobile.

* *Remembrance gathered from a folklore student, spring 2015, played as sixteen-year-old, Chesterton, Indiana; remembrance gathered from a folklore student, spring 2015, played as fourteen-year old at Disneyland, San Rafael, California.*

B5. Hole in Your Hand

The actor holds up a tube, like a paper-towel tube or rolled-up magazine, to his eye, then holds up his other hand in front of his other eye, next to the tube, with palm facing his face and both eyes open. If the actor focuses on an object in the distance, he will see it through the palm of his hand, as if the tube were piercing it.

* *Observation of Will Rice, spring 2012, fifteen years old, Lafayette, Louisiana; see also Colman's report of the first appearance of the illusion in print in 1871 (2015, 343).*

B6. Mirror Summoning

In children's folklore, the mirror apparition is frequently cited—in one form or another—as a ritualesque and spiritual performance of preadolescence. During mirror apparitions, one or more youths enter a darkened space in front of a mirror. While gazing into the darkened mirror, the actors chant a traditionalized verbal component meant to bring forth a paranormal, usually malicious entity. We deal directly—and explicitly—with mirror apparitions in chapter 6 because traditions like Bloody Mary, Candyman, and I Hate the Bell Witch push the boundaries of the analytical genre that is folk illusions.

* *For folkloristic studies, see Dundes (1998), Langlois ([1978] 1980), and Tucker (2005). For studies on parallel experiences induced in scientists' laboratories, see Schwarz and Fjeld (1968) and Caputo (2010).*

B6A. BLOODY MARY

One or more actors stand in front of a darkened bathroom mirror and recite the name "Bloody Mary" a number of times (in our survey work, the number *three* is most frequently reported). After the verbal performance, the figure of Bloody Mary is visually perceived by one or more of the actors.

* *Remembrance gathered from a folklore student, fall 2013, played as eight- to nine-year-old, Chicago, Illinois; remembrance gathered from a folklore student, fall 2013, played as eleven-year-old, Evergreen, Colorado; remembrance gathered from a folklore student, fall 2013, played in sixth grade at a slumber party, Michigan City, Indiana.*

B6B. I HATE THE BELL WITCH

Best known in middle Tennessee, a place where legends circulate of a witch who haunted the Bell family of Adams, Tennessee, from the years 1817 to 1820. To coax the witch's apparition, one or more actors gaze into a darkened bathroom mirror and recite the phrase "I hate the Bell witch" an allotted amount of times. The witch is reported to be seen in the mirror, at which point she scratches the face or back of one or more performers.

* *Remembrance gathered from Nik Norrod, spring 2015, played in middle school, Adams, Tennessee; remembrance gathered from Addy Harris, fall 2014, played in elementary-school bathrooms, White House, Tennessee; for folkloristic accounts of the legend, see Hudson and McCarter (1934) and Lockhart ([1984] 2009).*

B6C. ELEANOR, ELEANOR, ELEANOR

A tradition localized to Wellington School in Columbus, Ohio, a performance of Eleanor, Eleanor, Eleanor consists of one or more actors standing in front of a darkened bathroom mirror. Then, the actor spins three times. During each spin, the actor speaks the name "Eleanor," a student who mysteriously died in front of one of the school's mirrors many years ago. Eleanor is reported to then appear in the mirror.

* *Remembrance gathered from Nora and Maggi Cashman (twelve and fourteen years old, respectively), fall 2015.*

B7. Afterimage

Arguably best categorized as a popular-culture illusion, some form of Afterimage is frequently mentioned alongside folk illusions. The most cited is the Jesus Afterimage illusion, which arises when an actor gazes at a specially designed black and white presentation of a face on a computer screen. The actor then closes her eyes and a negative afterimage of Jesus's face appears in the closed-eye view.

* *Remembrance gathered from a folklore student, spring 2014, played as ten-year-old, Deerfield, Illinois; remembrance gathered from a folklore student, spring 2015, played as sixteen-year-old, Elkhart, Indiana.*

B8. Hand on Your Shoulder

The agent asks the patient whether or not the patient believes in ghosts. Usually the patient demurs. The agent then asks, "Well, then, what is that hand on your shoulder?" When the patient turns to look, all of a sudden, she sees a vague hand on her shoulder.

* *Remembrance gathered from eighteen-year-old Danish student in Lafayette, Louisiana, fall 2012.*

B9. Missing Finger

The agent's hands are interlocked and folded as in prayer, but the ring finger of one hand is slid underneath the ring finger of the other. The "top" ring finger covers the hidden finger's knuckle, so the folded hands appear to be missing one of the ten fingers.

* *Remembrance gathered from Corey Green, age thirty-three, and Jennifer Pecora, age twenty-six, fall 2010; see also Welsch (1966).*

B10. Disappear

Two actors stand across from one another on camping trips or summer nights at a distance of about twenty feet. Without moving, the two actors stare at one another. After a priming period of about twenty seconds, both actors start to turn to gray and eventually disappear from the other actor's view.

* *Remembrance gathered from a folklore student, spring 2015, played from the ages of ten to eighteen, Arlington, Virginia.*

B11. A Growing Dot

An actor draws a dot on a sheet of paper. Then, the actor stares at the dot for an appropriate priming period, after which the dot begins to grow.

* *Remembrance gathered from a folklore student, spring 2015, played in primary school, Tangshan, Hebei (China).*

B12. Snake in the Cooler

The director instructs the actor to retrieve an item from a cooler (e.g., a beverage or a bag of ice). When the actor goes to the cooler and opens the lid, a rubber snake, which has been attached via a fishing line to the lid of the cooler (one end of the line attached to the cooler and one end of the line attached around the rubber snake's head), rises as the actor lifts the lid. The illusion is successful if the actor is startled by the rubber snake.

* *Observed by Barker in summer 2013, Pensacola, Florida; see chap. 1.*

B13. Floating Finger

(See Video 9.) An actor holds her index fingers in front of her face so that the tips of the index fingers are pointed at each other. Then, the actor stares into the open space between the fingertips and slowly begins to move the fingertips together. At a certain point, the eyes begin to cross and a floating third index finger appears in between the fingertips, which still are not touching. One remembrance describes the floating finger as a "hot dog."

* *Remembrance gathered from a folklore student, fall 2013, played between the ages of five and eight, Granger, Indiana; remembrance gathered from a folklore student, spring 2014, played in fourth grade, Wabash, Indiana.*

B14. Zane's Illusion

(See Video 10.) The actor lightly presses her right palm against her forehead so that her wrist and/or the distal portion of her forearm rest against the space between her eyes (her glabella). Next, the actor forms her left hand into a standard pointing position. Then, while looking straight ahead, the actor moves her left index finger from the left to the right side of her visual field. The pointed index finger should pass behind the right forearm. The forearm's obstruction of her binocular vision creates the visual illusion that her pointing finger is shrinking or even disappearing.

Remembrance gathered from Zane Hillenburg, age ten, spring 2017, Bloomington, Indiana (see the preface); remembrance gathered from Seamus McDonald, age twelve, fall 2018, Bloomington, Indiana.

C. Auditory Illusions

C1. Buzzing Bee

(See Video 11.) The agent wets the tip of her index finger. Next the agent—often sneakily—moves that finger next to the patient's ear. Finally, the agent rubs her wet fingertip against her thumb, making the sound of a buzzing bee.

* *Remembrance gathered from Ellen Hesse, fall 2011, played in middle school, Lafayette, Louisiana.*

C2. The Church Bell

(See Video 12.) The actor gets a piece of thread, twine, or other filament about four feet (fifty centimeters) in length. The actor holds one end in each hand, then hangs a coat hanger on the thread, wrapping the thread around the hanger to secure it. She then wraps each end of the thread several times around the correlative index finger and places a finger in each ear. Moving her torso back and forth to swing the coat hanger a bit, she strikes the hanger against any firm object, like a chair or door frame, preferably one that is metal. Alternatively, the director can strike the hanger with a spoon. Observers will only hear a light sound, but the actor will hear a surprisingly loud ringing sound, often described as a church bell. This is also sometimes called the Coat Hanger Bell or the Church Bell illusion.

* *Remembrance gathered from a folklore student, spring 2015, played as eight-year-old, Vincennes, Indiana; see also Rohault, who first described the illusion in his* Treatise on Physics *as "an experiment which is often made use of to divert Children" ([1671] 1723, 183).*

C3. Ocean in a Shell

A widely recognized folk illusion in which the actor holds a conch shell up to her ear. The actor hears the sound of the ocean in the conch shell.

* *Remembrance gathered from a folklore student, spring 2015, played in preschool, Vincennes, Indiana.*

The illusions C4–C12 constitute a recognizable subcategory we refer to as audiovisual miming illusions (note: C11 has an important haptic quality as well). In these forms, youths demonstrate their awareness of the power of multisensory perception, especially as it relates to the coalescence of culturally relevant mimed actions with perfectly timed acoustic elements that are commonly associated with those actions. We discussed at length Barker's experience of being duped as an unsuspecting patient by a performance of Ping-Pong at a Boy Scout camp in chapter 3, and it is important to note here that several of these audiovisual miming illusions—especially Haircut, the Ripper, and Cracked Nose—are performed with the intention of tricking an unsuspecting patient into perceiving an undesirable or grotesque action.

Simultaneity is known to play an important role in multisensory perception. The best-known examples are simultaneous visual and haptic stimuli during out-of-body perceptions, such as the rubber hand experiment and the ventriloquism effect.

C4. Haircut

The agent makes a show of cutting scissors behind the patient's head. Even if the agent tells the patient that she is not really going to cut the patient's hair, the patient still feels as though her hair is being cut because the scissors "sound like they are close to your head." The scissors do not have to be close to the patient's head (usually the scissors are one or two feet away), but the sound "will usually trigger the same reaction."

* *Remembrance gathered from a folklore student, spring 2015, played in middle school, Annandale, Virginia.*

C5. The Ripper

The agent snatches a piece of important paper from the patient (e.g., the patient's homework, a note from the patient's teacher, or a drawing the patient is working on); then, the agent holds the paper on one end in between his index finger and thumb. The agent makes a ripping motion with one hand down the paper while blowing over the paper in such a way as to mimic the sound of ripping paper.

* *Remembrance gathered from Allen Jones, fall 2011, Northern California.*

C6. Ping-Pong

The directors perform an illusory match of Ping-Pong without the use of a ball by timing pantomimed swings with the flicking of a Styrofoam cup. The flicking sound is timed with the precise moment when an actual ball would collide with a paddle. The back and forth of this illusory match is observed by an audience of actors. The actors' heads follow the flight of the imaginary ball. In performances that involve a naive patient, the patient experiences the illusion that the entire match (including the actions of the directors and actors) is an actual Ping-Pong match with an actual ball flying back and forth.

* *Observation of Boy Scouts at Lost Bayou Scout Camp in Washington, Louisiana, summer 2012.*

C6A. PING-PONG (STOLEN BALL)

An interrupted version of Ping-Pong involves one of the actors stepping into the path of the flying illusory ball while pantomiming actions correlative to taking the ball. The illusory ball is then kept away from the directors, so that they are unable to continue the match.

* *Remembrance gathered from Brandon Dale, summer 2012, played as Boy Scout at camp, Lafayette, Louisiana.*

C6B. SNAP TOSS

In the Snap Toss variation, two directors mimic athletic games in which a ball or object is thrown back and forth by snapping their fingers at the precise moment that the "object" leaves the throwing director's hand *and* similarly snapping the fingers at the precise moment that the "object" lands in the catching director's hand.

* *Remembrance gathered from a folklore student, played during class in middle school, Floyds Knobs, Indiana.*

C7. Cracked Nose

The agent places both of his hands over his nose and mouth, in the same position that frequently accompanies a sneeze or cough. Then, the agent shifts his hands to one side of his face without uncovering his mouth or nose. As he does so, he simultaneously bites down on uncooked pasta which, for the unsuspecting patient, creates the illusion that the agent has cracked his nose.

* *Remembrance gathered from Dominik Tartaglia, fall 2013, played as fourth grader, Toledo, Ohio; remembrance gathered from Will Rice, age fifteen, spring 2012, who snapped his thumbnail against his upper teeth to create the noise.*

C8. Sousaphone Lightsaber

Two directors, members of a school or marching band, who play the sousaphone (a kind of tuba worn around the neck in order to facilitate marching), face one another with toy lightsabers (the science-fiction swords associated with the very famous *Star Wars* films). The directors then begin to mime the actions of a lightsaber duel while playing low, uneven tones that mimic the famous sounds of lightsabers.

* *This particular folk illusion was brought to our attention by Joel Chapman in spring 2016. He cites an example that can be viewed online (see EvanCG Productions 2014).*

C8A. PICK-A-NOTE LIGHTSABER

"A chaotic band activity where every student picks a note, usually one that's within a fourth (perfect fourth example, C could play up to F, down to G—including any sharp or flat). Then, the conductor points to different sections or individuals to play louder. The dynamic cues from the conductor, and the clash created by half-steps and whole steps creates the tense sound that simulates a lightsaber."

* *Remembrance gathered from Joel Chapman, spring 2016, played in high school, Indiana.*

C9. Blow Dart

The director mimes the position and actions of shooting a blowgun and simultaneously produces a loud blowing noise like that of shooting a blowgun. The actor who is "hit" falls to the ground and cannot move until some other participant mimes pulling the blow dart out.

Remembrance gathered from a folklore student, spring 2014, played in high school, Wheaton, Illinois.

C10. Loud Slap (or Punch)

Another frequently mentioned audiovisual miming illusion, the director in Loud Slap swings with one hand, miming a slap to the actor's face. Simultaneous with the precise moment that the director's hand passes (without striking) the actor's face, the director slaps his own chest. If properly timed and performed, unsuspecting patients can be tricked into believing the illusory perception of an actual slap.

* *Remembrance gathered from a folklore student, spring 2014, played in grade school, Cincinnati, Ohio.*

C11. Spit-Wad

The agent stands behind the patient and snorts through his nose as if gathering mucus in his mouth. The agent then blows a forceful puff of air into the patient's hair. The patient feels as though the agent has spit a wad of mucus into the patient's hair.

* *Remembrance gathered from Corey Green, fall 2011, age thirty-three, Bee Branch, Arkansas.*

C12. Guillotine

From Simon J. Bronner: "One day a troop of Boy Scouts were initiating some new members. One of the initiates was blindfolded and placed with his head in a guillotine. A noise was made which sounded like the blade of the guillotine being released, and at the same time a wet towel was slapped across the boy's neck. The troop members screamed to make the initiate think that something had gone wrong. The initiate, hearing the screams and feeling the towel, was shocked."

* *See Bronner (1988, 169).*

D. Motor Illusions

D1. Winding-Cranking Fingers

(See Video 13.) The actor's hands and fingers are loosely twisted together as in prayer *except* for the index fingers, which are pointing up and parallel, not touching one another. The director pretends that there is a crank above the index fingers and begins to turn that crank; the director must verbalize these two components of Cranking Fingers. As the index fingers tend to relax and come closer together, the actor feels as though the director is *cranking* his index fingers closer together.

* *Remembrance gathered from a folklore student, spring 2015, played in grade school, Merrillville, Indiana; remembrance gathered from a folklore student, spring 2014, played at nine or ten years old (cranking described as a fishing-reel motion, Carmel, Indiana; observation of seventh-grade boys and girls at St. Cecilia Middle School, spring 2011, Broussard, Louisiana (rather than a praying position, the cranked index fingers were held out in front of the actor's body, parallel to the ground).*

D1A. TIED-UP FINGERS

The director and actor perform D1 except that the director mimes tying an invisible string around the actor's index fingers. The tightening of this "string" brings the index fingers together.

* *Remembrance gathered from a folklore student, played with friends as a seven-year-old, Borden, Indiana.*

D1B. MAGNETIC FINGERTIPS

The actor assumes the same performance position as D1 and D1a. As the actor's index fingers relax, they are perceived to be pulled together by magnetic force.

* *Remembrance gathered from a folklore student, spring 2015, performed at camp as ten- or eleven-year-old, West Windsor, New Jersey.*

D2. Magnetic Rocks

(See Video 14.) Magic Rocks and its variants constitute another set of traditionalized Kohnstamm illusions. The actor is told to place a rock in each hand. He is then instructed to press the rocks together very firmly. An incantation is performed by the director; this incantation allows for an allotted amount of time to pass. When the rocks are *slowly* pulled apart, the actor feels as though they are magnetic.

* *Remembrance gathered from a folklore student, spring 2011, played as a child, Memphis, Tennessee; remembrance gathered from a folklore student, fall 2013, played with brothers around eight years old, Fairfield, Connecticut.*

D2A. MAGNETIC PENCILS

The actor completes the same kinesthetic activities as D2 except the actor presses together the rubber end of pencils for a priming period of ten seconds. No incantation was reported. When the actor slowly pulls apart the pencils, they feel as though they have been magnetized.

* *Remembrance gathered from a folklore student, spring 2014, played as fifteen-year-old, Vietnam.*

D2B. MAGNETIC FISTS

The actor completes the same actions as D2 and D2a except the actor holds no objects. Instead, she presses together the knuckles of her balled fists. At the conclusion of the priming period, the fists feel as though they are magnets attracting one another.

* *Remembrance gathered from a folklore student, spring 2015, played as nine- to ten-year-old, Pune, Maharashtra (India); remembrance gathered from a folklore student, spring 2015, played as eight-year-old by "grinding knuckles together," Mumbai, Maharashtra (India); remembrance gathered from a folklore student, spring 2014, played with cousins at grandmother's house around ten or eleven years old, Guilford, Indiana.*

D3. Rubber-Band Hands

The actor completes the actions of D2b—pressing the knuckles of her fists together for a priming period of approximately thirty seconds. While slowly pulling apart her fists, the actor reports that her fists are being held together by invisible rubber bands.

* *Remembrance gathered from a folklore student, spring 2014, played as third grader, Plainfield, Indiana.*

D4. Rod and Reel

The actor makes a fist with each hand and presses the thumb sides of the fists together. Then, he moves the left hand like he is reeling in a fishing rod. After doing this for thirty seconds, "it will be tough to move the fists apart."

* *Remembrance gathered from a folklore student, spring 2014, played as sixth grader in middle school, Indianapolis, Indiana.*

D5. Double-Magnetic Fists

Two actors make a fist. Then the two actors push the knuckle sides of their fists against one another for approximately thirty seconds. When the actors stop pushing and try to pull their fists apart, they feel as though a magnetic force keeps them from doing so.

* *Remembrance gathered from a folklore student, spring 2014, played during lunch and recess in second or third grade, Springfield, Virginia.*

D6. Frozen Fist

The actor is instructed to grasp the index finger of the director (or another person) and squeeze as tightly as possible for a count of 250. The grasp is then loosened *slightly* so that the index finger can be slipped out. The director rubs a quarter or fifty-cent piece over the partially closed hand. A freezing sensation is felt. Difficulty is experienced in opening the hand.

* See Welsch (1966, 177) for coin variant; remembrance gathered from a folklore student, spring 2014, played as eight-year-old, Glenview, Illinois.

D6A. FROZEN FINGERS

The director instructs the actor to hold a pencil very tightly between his index finger and thumb. After a priming period of thirty seconds, the director removes the pencil, and the actor feels as though he cannot move his fingers, as if they are frozen.

* Remembrance gathered from a folklore student, spring 2015, played as eight- to twelve-year-old, Marietta, Georgia.

D7. Tied-Up Fists

The director instructs the actor to make a tight double-fist with both hands. For a priming period of about thirty seconds, the director mimes the actions of tying a string or rope around the actor's fists. Then, the actor tries to slowly open his fist. Opening the fists is surprisingly difficult.

* Remembrance gathered from a folklore student, fall 2013, played around ten years old the first time and several times since, Chicago, Illinois; remembrance gathered from a folklore student, spring 2016, played in middle school with the addition of the actor holding pencils in her clenched fists, Crown Point, Indiana.

D8. Floating Arms

The director instructs the actor to place her arms at her sides with her palms touching her sides. The director then holds the actor's arms down as the actor tries to lift them from her sides for approximately thirty seconds. The director then removes her hands, and the actor's arms feel as though they are floating at her sides.

* Observation of seventh-grade girls at St. Cecilia Middle School, 2011, Broussard, Louisiana. This is the second best-known folk illusion in our surveys. In a survey of 227 students at Indiana University in the spring of 2016, 55.07 percent of respondents reported playing some version of Floating Arms. Floating Arms' associated psychophysical process, the Kohnstamm phenomenon, is widely considered in the experimental literature. See Kohnstamm (1915), Craske and Craske (1986), Ivanenko et al. (2006), Ramachandran and Ramachandran (2008), Duclos et al. (2007), and De Havas et al. (2016).

D8A. FLOATING ARMS (DOOR FRAME)

The director instructs the actor to press the outside of her arms against a door frame for approximately thirty seconds. When the actor steps away from the door frame, her arms feel as though they are floating.

* Remembrance gathered from a folklore student, spring 2014, played with brother in elementary school, Fishers, Indiana; remembrance gathered from a folklore student, fall 2013, played as ten-year-old, Ann Arbor, Michigan.

D8b. Floating Arms (Pockets)

The director instructs the actor to place his hands in his own pants pockets. The actor then tries to lift his arms out and up (against the inside of his pants' pockets) for about thirty seconds. The actor then removes his arms. His arms feel as though they are floating at his sides.

* Observation of Will Rice, spring 2009, age twelve, Lafayette, Louisiana; remembrance gathered from a folklore student, spring 2015, played as eight- to twelve-year-old at school, Cherry Hill, New Jersey.

D8C. FLOATING ARMS (POP THE BUBBLES)

The director and actor perform the kinesthetic actions of Floating Arms as in D8. As the actor's arms are rising, the director mimes "popping the bubble" that is "lifting" the arms. As the director does this, the actor's perception that her arms are rising stops.

* Remembrance gathered from a folklore student, spring 2015, played in class as a sixth grader, Fairhaven, New Jersey.

D9. Frankenstein's Arms

The director and actor perform the kinesthetic actions of Floating Arms as in D7; however, when the director releases the actor's arms, the director "pretends to crank them [the actor's arms] up like Frankenstein."

* Remembrance gathered from a folklore student, spring 2015, performed at approximately ten years old, Cleveland, Ohio.

D10. Devil's Arm Lift

The director tells a story about the devil. The director then snaps his fingers right next to the actor's ear, and the actor extends his arms and crosses them ten times. Repeat this process ten times. After completing the process, the actor says, "I wish my arms would rise," and they rise.

* Remembrance gathered from a folklore student, spring 2014, played as ten-year-old, Querétaro, Mexico. Additionally, though we have yet to gather a remembrance from someone who has participated in the YouTube tradition of "devil lifts my arms challenge," readers are advised to search those terms. Searching "devil lifts my arms challenge" in September 2016 brought up more than twenty recordings of youths framing a Kohnstamm experience with the belief that the devil was lifting their arms.

D11. Force between Your Hands

The actor's hands are held together in front of her chest. Clasping the hands tightly, she tries to pull them apart against the force of her own grip for a priming period of about thirty seconds. At the conclusion of the priming period, the hands are separated and the actor experiences difficulty pressing them back together.

* *Remembrance gathered from a folklore student, fall 2013, usually played at school between eight and ten years old, Chicago, Illinois.*

D12. The Invisible Ball

The director instructs the actor to hold her hands out in front of her sternum as though she is holding an imaginary soccer ball. Then, the director places her hands on the outside of the actor's hands. The director presses forcefully against the outside of the actor's hands (as though to bring the actor's hands together), and the actor presses forcefully outward for a priming period of about thirty seconds. Then, the actor's hands are released and she perceives an invisible ball between her hands.

* *Remembrance gathered from a folklore student, spring 2014, played as ten- to thirteen-year-old, Manalapan, New Jersey; remembrance gathered from a folklore student, spring 2015, played in elementary school, Columbus, Ohio; remembrance gathered from a folklore student, spring 2015, played as eight- or nine-year-old and described the sensation as that of a "bubble" between the hands, Delray Beach, Florida.*

E. Body Schema Illusions

The term *schema* (including *body schema*) has come to have very specific—and not necessarily identical—meanings in different fields of cognitive and physiological science (e.g., perception, development, and motor skills); for a brief summary of these issues, see Arbib (2004). In reference to the folk illusions cataloged in this section, we intend a more inclusive and general definition of body schema as the active perception of the body's (com)position and capabilities in a given spatial context.

Some of the folk illusions, such as Can You Kiss Your Elbow?, do not—necessarily—involve a perceptual illusion. We include them, instead, as examples of cognitive illusions that are based in underrefined or incorrect knowledge and beliefs about the body. Other folk illusions in the Body Schema category, such as Got Your Nose, very well may include a perceptual illusion, where haptic aftereffects of having one's nose pinched are misperceived as the sensation of the absence of the patient's nose.

E1. Got Your Nose

The agent curls her index and middle fingers slightly, then uses the two fingers as pinchers to grab the patient's nose. The agent stands back while surreptitiously slipping the tip of that hand's thumb in between the two fingers. The agent then tells the patient "I've got your nose" while showing the patient the two fingers with the thumb tip sticking out. The visual presentation of the *thumb nose* and the haptic aftereffect of having one's nose pinched prompts an inquisitive illusion for young patients who may believe that their noses have actually been snatched away!

* *This is a widely recognized folk illusion. In the summer of 2014, we witnessed a poignant and playful performance of this folk illusion in Bloomington, Indiana, between a preschool teacher and her student after which the three-year-old patient felt his face multiple times.*

E2. Growing/Shrinking Arms

The actor stretches his arm so that it is just long enough to touch the wall. He then brings the hand of that arm up to its correlative shoulder. Then, the actor strikes that bent elbow several times. When the actor once again reaches for the wall, his arm will be noticeably shorter.

* *Observation of seventh graders at St. Cecilia Middle School, spring 2011, Broussard, Louisiana.*

E2A. GROWING/SHRINKING ARMS VARIANT

The actor puts his arms out, parallel to the floor so that his hands are pressed together in a "diving" position. In this beginning performance position, his arms will be equal length. He then stretches one of his arms and quickly slaps them back into the original performance position. The stretched arm will be longer.

* *Observation at St. Cecilia Middle School, spring 2011, Broussard, Louisiana.*

E2B. GROWING/SHRINKING ARMS VARIANT

The actor stretches both arms out, palms down, in front of himself. He notes both arms are the same length. The actor then bends one arm up, touching his shoulder with his fingers. He then rubs his elbow a few times with his other hand, then extends arms again. It is noticed that the rubbed arm is shorter than the other arm.

* *Remembrance gathered in Lafayette, Louisiana, 2012.*

E3. I Can't Move My Finger

(See Video 15.) The actor places his hand and fingers on a flat surface so that the finger pads of the thumb, pointer, ring, and pinky are on the table surface. The middle finger, though, is bent at the second knuckle so that the knuckle and the top half of the finger are resting on the surface. In this position, the ring finger becomes "paralyzed" and is very difficult for the actor to lift. Alternatively, the actor may place both of his hands (in the same shape described above) together so that the finger pads of the thumb, pointer, ring, and pinky are touching each other while the second knuckles of the middle finger touch each other. In this case, actors will find moving either ring finger to be very difficult.

* *Remembrance gathered from a folklore student, fall 2013, played as twelve-year-old, Maryland; remembrance gathered from a folklore student, played at sleepovers as eight-to ten-year-old, Mt. Prospect, Illinois; for the two-hand variant, see Welsch (1966, 177).*

E4. Twisted Hands

(See Video 16.) This is called "the Japanese illusion" in the human physiology literature (see Van Riper 1935). The director instructs the actor to cross his hands at the wrist, palms down. The palms are brought together and the fingers are folded together. The hands are brought down beneath the wrists and up before the chest. The fingers now lie directly beneath the eyes with the thumbs pointing away from the body. The director then points to various fingers of the folded hands but does not touch them. The actor is challenged to wiggle the finger to which the director has pointed. It is surprisingly difficult.

* *In classroom surveys administered in 2008–18, we gathered remembrances from university students who performed Twisted Hands beginning as early as seven years old from Illinois, Indiana, Louisiana, California, Michigan, Ohio, and New York. We also received remembrances from students who grew up in Zhejian, Si Chuan, and Jiangso, China, and the United Arab Emirates. See also Welsch (1966, 176), Arnold (1952), Boulware (1951), and Klein and Schilder (1929).*

E4A. TWISTED HANDS (FUNNY FACE)

The actor assumes the position of Twisted Hands. Then, bringing her twisted hands close to her face, she places her right index finger on the right side of her nose and her left index finger on the left side of her nose. Because her hands are crossed, it is difficult to decipher which finger is which, but if the actor has done so correctly, she can untwist her hands while keeping her index fingers on her nose. Then, she wiggles her fingers in front of her face.

* *Observation of four twelve-year-old girls, summer 2009, Lafayette, Louisiana; remembrance gathered from Prasanna (Monty) Kawatkar, summer 2009, played as a child in Mumbai, India; remembrance gathered from a folklore student, spring 2015, played as seven-year-old on playground, Liaoning, China.*

> Folk illusions E5–E8 focus on play with movement and coordination. We include them in the category of body schema illusions because they seem to play with children's expectations of what they can or cannot do. Tradition provides impetus for trying to rub one's head while patting one's belly, and, within the tradition, children play with their expectations about the ability to succeed at a pair of seemingly simple tasks—patting and rubbing. Failure produces surprise and—interestingly—fun. We also notice that these coordination illusions have no narrative frame outside of the local performance frame itself. Generally, they feature no verbal characterizations or explanations. People seem to know that bodies can sometimes fail to coordinate smoothly even when there is no immediately clear reason for them to do so.

E5. Pat and Rub

A classic children's activity in which the actor pats her head while trying to rub her belly. The illusion arises from conflict between the expected level of difficulty associated with this coordination trick and the actual level of difficulty. Patting one's head and rubbing one's belly at the same time is generally more difficult than expected.

* *Observation of six four- to eight-year-old boys and girls, summer 2014, Bloomington, Indiana.*

E6. Nose and Ear

An actor alternates as fast as possible his hands, which are pinching the nose and the ear (while his arms are twisted and untwisted and retwisted). The activity is performed with an accompanying tune that the performer hums.

* *Remembrance gathered from Cory Green, fall 2010, age thirty-three, Bee Branch, Arkansas.*

E7. Finger Twirl

The actor points the tips of her index fingers toward one another in front of her sternum. Then, the actor attempts to twirl index fingers in opposite directions. Alternatively, the actor may try to twirl closed fists in a similar fashion.

* *Observation at St. Cecilia Middle School, spring 2011, Broussard, Louisiana; remembrance gathered from a folklore student, spring 2016, played as twelve-year-old often while by himself, Indianapolis, Indiana.*

E8. Impossible Signature

The actor is instructed while sitting to move her foot in a circle, rotating from the knee. After a rhythm is established, the actor is instructed to write her signature in the air with her hand. When the actor starts to write, the foot deviates from the established circle, often switching involuntarily to tracing part of the signature also.

Remembrance gathered from Hannah Ritorto, age thirty-three, who played it in the doctor's office in middle school, Northern Virginia. See also Cobb (1999, 18).

E9. Can You Kiss Your Elbow?

A director asks a person, usually a child, if she can kiss (or lick) her own elbow. The child imagines she can and tries to do so, often distorting her body in the effort to succeed.

Remembrance gathered from Jaron Nazaroff, summer 2016, twenty-four, who played it "as a kid" in Santa Clara, California; observation of six four-to-eight-year-old boys and girls, summer 2014, Bloomington, Indiana.

F. Weight Illusions

In the category of weight illusions, we include folk illusions that play with the weight of an external object (F1–F3) as well as folk illusions that play with the felt weight of the experiencer's own body or body parts (F4–F6). It is important to note that we have not categorized all folk illusions that play with the ideas of weight in this category. We categorized Floating Arms (D8), for example, with other Kohnstamm activities in motor illusions. Likewise, we categorized Falling through the Floor in falling and flying illusions.

F1. The Not-So-Heavy Object

The agent approaches the actor carrying a seemingly heavy object—like a seemingly heavy box. The agent makes the object seem heavy to the patient by physically acting as though he is straining to hold the object. The agent then passes the seemingly heavy object to the patient, who is surprised to find that the object is not heavy—that the object is, for example, an empty box.

* *See Flanagan et al. (2001).*

F1A. THE NOT-SO-HEAVY HAMMER

The unsuspecting actor lifts a sledgehammer at the request of a tricky director. The hammer, though visually identical to a real sledgehammer, is actually made of lightweight wood and a drilled-out handle. As the actor lifts the "hammer," onlookers enjoy watching the actor's experience of the positional overshoot.

* *See the discussion of this sledgehammer illusion in chap. 5.*

F2. Light as a Feather

Often performed during sleepovers or in darkened spaces, four to six lifters sit around a patient who is lying on her back in a dimly lit room. The directing lifter—seated at the head of the patient—may tell a narrative about the death of the patient; then, all lifters chant the phrase "light as a feather, stiff as a board" several times. They then lift the patient off of the ground using only one or two fingers from each hand.

* *In classroom surveys administered in 2008–16, we gathered remembrances from university students who performed Light as a Feather beginning as early as six years old from Virginia, Indiana, Louisiana, California, Illinois, New Jersey, Colorado, Kentucky, New York, Georgia, Texas, Pennsylvania, and Connecticut. See also Opie and Opie (1959) and Tucker (2008).*

F3. Lift from a Chair

Three to four acting lifters lift a patient, who is seated in a chair. After a ritualesque/pseudoscientific prime, the second lift goes much higher than the first.

* *See Tucker (1984, 2007, 2008).*

F3A. COMBINE OUR STRENGTHS

Three to four acting lifters gather around the patient, who is seated in a chair. The lifters—using only two fingers on each hand—lift the patient out of the chair, marking the height of the lift. The lifters then stack their hands alternately above the patient's head (sometimes making sure not to touch hands) and lift again. This second lift is higher.

* *Observation of seventh-grade boys at St. Cecilia Middle School, spring 2001, Broussard, Louisiana.*

F3B. CUT THE CENTER OF GRAVITY

Three to four acting lifters gather around the patient, who is seated in a chair. The lifters—using only two fingers on each hand—lift the patient and the chair, marking the height of the lift. The lifters then stack their hands above the patient's head and recite a ritual incantation. Then, they lift again. This second lift is higher because the ritual components "cut the center of gravity."

* *Remembrance gathered from Josue, summer 2014, age twenty-seven, played as a youth in Venezuela.*

F4. Heavy Arm

The doctor (i.e., the agent) performs "surgery" on the patient's arm. The agent starts by running her finger up the patient's arm to "cut open" the arm. Then, the agent twists the skin of the arm and tells the patient that she is filling the arm with lead. Finally, the doctor pinches the skin up the arm where the "cut" was made, thus giving the patient stitches. Finally, the agent tells the patient to hold up both of her arms to see which one is heavier. The patient's arm that now has lead in it feels heavier.

* *Remembrance gathered from a folklore student, spring 2015, played in second or third grade at elementary school, Columbus, Indiana.*

F5. Buckets of Water

The actor mimes holding two buckets of water out to her sides with arms extended and parallel to the ground. The director then mimes pouring water or ladling water out of the buckets. The actor's arms rise and lower depending on the amount of "water" in the buckets.

* *Remembrance gathered from a folklore student, spring 2014, played as eight- or ten-year-old, Lancaster, Pennsylvania.*

F6. Sandman

The patient lies on her back on the floor while the agent sits at her side or at her head. The agent then begins to tell a story about the Sandman who has broken into the patient's house. In the story, the Sandman stomps up the stairs, and as the agent performs this verbal component she also pounds on the ground around the patient's body—mimicking the Sandman's stomps. Then, the Sandman knocks the patient out (the agent lightly hits the patient on the head). The Sandman cuts the patient open (the agent runs her fingernail over the patient's forehead) and begins to fill the patient with sand (the agent presses firmly all over the patient's body). After repeating the sand-filling portions of the activity about four times, the patient is instructed to try to get up, but getting up is difficult because the patient feels heavy.

* *Remembrance gathered from a folklore student, spring 2014, played at ten to thirteen years old at sleepovers, Fishers, Indiana; remembrance gathered from a folklore student, fall 2013, played at an overnight summer camp while in middle school, Deerfield, Illinois; remembrance gathered from a folklore student, spring 2016, played as ten-year-old, Oklahoma City, Oklahoma.*

G. Falling and Flying Illusions

G1. Falling through the Floor

(See Video 17.) The actor lies on his belly while the top half of his body is held off of the floor by the director, who is holding on to the actor's wrists. This continues for thirty or more seconds. When the actor is slowly lowered to the floor, he will have the sensation that he is falling through the floor.

* *Observation at St. Cecilia Middle School, spring 2011, Broussard, Louisiana; remembrance gathered from a folklore student, spring 2014, played in middle school, Greencastle, Indiana; remembrance gathered from a folklore student, spring 2015, played as freshman in high school, Moken, Illinois.*

G1A. FALLING THROUGH THE FLOOR (FREEFALLING)

Same as G1, but the illusion is a sensation of "freefalling."

* *Remembrance gathered from a folklore student, spring 2014, played as eleven-year-old, Fort Wayne, Indiana.*

G1B. FALLING THROUGH THE FLOOR (FALLING OFF A BUILDING)

Same as G1 and G1a, but the illusion is a sensation of "falling off of a building."

* *Remembrance gathered from a folklore student, spring 2014, played at ten to twelve years old, Princeton, New Jersey.*

G1C. FALLING THROUGH THE FLOOR (BURIED IN A GRAVE)

The kinesthetic actions and performance positions are the same as G1, but during the priming period in a performance of Buried in a Grave, the director tells a narrative of how the actor has died, and when the actor is lowered, she feels as though she is going underground "into her grave."

* *Remembrance gathered from a folklore student, spring 2014, played from eight to twelve years old, Sacramento, California.*

G2. Flying

The kinesthetic actions and performance positions of Flying are the same as G1; however, the created illusion was reported as a "flying sensation."

* *Remembrance gathered from a folklore student, spring 2014, played at camp while in grade school, New York.*

G3. Legs through the Floor

The actor lies on her back and lifts her legs up from the floor (to about a forty-five-degree angle). The director holds her legs in this lifted position for approximately forty-five seconds, then lowers them slowly to the floor. The actor feels as though her legs are falling through the floor.

* *Remembrance gathered from a folklore student, spring 2014, played in middle school, South Korea; remembrance gathered from a folklore student, fall 2013, played as eight-to twelve-year-old, Columbus, Ohio; remembrance gathered from a folklore student, spring 2015, played at sports practice in high school, Indianapolis, Indiana.*

G4. The Lean

The actor leans against a chair (which is leaned against a wall at a forty-five-degree angle) for an allotted amount of time; when the actor stands up and walks, he feels as though he is leaning forward.

* *Observation of Will Rice, fall 2010, twelve years old, Lafayette, Louisiana.*

G5. Cliffhanger

The agent stands behind the patient and tells the patient that she is standing at the edge of a cliff. The agent instructs the actor to close her eyes and look down (over the edge). The agent then suddenly shoves the patient in the back. Afterward, the patient is asked what color she saw when she was shoved. This color is said to correlate with where (or into what) the actor fell: black = death, white = heaven, green = grass, blue = ocean, etc. We have also received remembrances that report Cliffhanger being performed as the concluding element of the Chills (A3).

* *Observation of four twelve-year-old girls, summer 2009, Lafayette, Louisiana; observations at St. Cecilia Middle School, spring 2011, Broussard, Louisiana.*

G6. "Lifted" Table

The actor stands on a table, or some other flat surface that can be lifted. Four lifting directors stand on the sides of the table. The actor is then told to close his eyes and put his hands on the heads of the lifters at his sides. Then, the lifters act as though they are lifting the table several feet off of the ground. They do this by lifting the table only inches off of the ground while slowly kneeling or bending lower so that the actor's arms must reach further downward in order to maintain contact with the lifting directors' heads. Once the table has reached a sufficient "height," the directors pretend as though they are losing strength of their grip on the table and force the actor to jump off. The actor, quite terrified by the oncoming fall, is surprised to learn that he has only been lifted a few inches from the ground. Alan McHugen (see below) remembers that the actor "lay on the floor shaking and sobbing."

* *Remembrance gathered from Mike Quinn, summer 2016, sixty years old (performed as a youth in Minnesota), Lafayette, Louisiana; remembrance gathered from Alan McHugen, summer 2016, witnessed performance as fifteen-year-old in 1969 at summer camp in Nova Scotia.*

G7. Mosh Pit

The actor stands still, locks up his body, and closes his eyes. Approximately six to eight directors stand in a tight circle around the actor and lightly push him back and forth, letting him repeatedly tip over in different directions. After some time, the actor feels "a sense of weightlessness and directional confusion."

* *Remembrance gathered from a folklore student, spring 2016, he and his friends did this activity in their middle school hallway in eighth grade, Vincennes, Indiana.*

H. Strength and Balance Illusions

H1. Salt Is Bad for Your Heart

The director informs the actor that "salt is bad for your heart." Salt is so bad that "just placing a packet of salt near your heart will weaken you." The director then instructs the actor to lift his left arm to the side, parallel to the ground. The director then tells the actor to stop the director from pushing the actor's raised arm down. On the first try, it is difficult for the director to push the actor's arm down. Then, the director has the participant place a salt packet in his right hand. The actor then places that hand containing the salt over his heart. Again, the actor lifts his left arm, and the director tries to push the arm down toward the ground. This time, the arm goes down easily, "proof that salt is bad for the heart."

* *Remembrance gathered from a folklore student, spring 2014, age thirty-seven, Lafayette, Indiana.*

H2. Magic Rocks

The director—who lets his audience know that he has "magical rocks"—instructs the actor to cup his hand next to his leg while standing straight up. The director would put his fist into the actor's cupped hand and pushes down. On the first try (without the rock), the actor easily and quickly loses his balance. On the second try, the actor is asked to hold the magic rock in his opposite hand. With the magic rock providing balance support, the director is not able to move the actor.

* *Remembrance gathered from Matt Brennan, spring 2015, played as ten- to fourteen-year-old, Fort Mitchell, Kentucky. Note: Matt adds, "The rocks turned into bracelets*

called 'Power Balance' by the time I went to high school and their power was shown in the same way as the rocks."

H3. You Can't Move My Hand

The director places his hand on the top of his head. He then challenges would-be actors by suggesting that no one will be strong enough to pull his hand from his head. Pulling the hand off of the director's head is surprisingly difficult. Our spring 2014 remembrance of the activity reports that "it was impossible to pull it off, no matter how hard they pulled."

* *Remembrance gathered from a folklore student, spring 2014, played on the playground in grade school, Palatine, Illinois.*

H4. Star Tipping

This illusion is performed outside at night. An actor stands and looks straight up at a star, then spins herself around until dizzy and the director yells "Stop!" The director then shines a flashlight in the actor's eyes, on which the actor becomes very disoriented, feeling that the light is a physical force that causes her to collapse to the ground.

* *Remembrance gathered from an English student, fall 2012, who played it in her late teens in Lafayette, Louisiana; remembrance gathered from a folklore student, spring 2016, who played it at age eighteen in Carmel, Indiana.*

I. Painful Illusions

I1. Doctor's Shot

In this activity, the agent explains that he will give or is in the act of giving a shot to the patient. The agent then pinches a large portion of the patient's flesh between the thumb and index finger of one hand, then suddenly jabs that flesh with a single protruding second knuckle on the middle finger. Counterintuitively, the jab yields a single sharp pain, feeling like a shot. The pinched flesh can also be pressed firmly with a finger to yield the sharp pain. The informant also noted that the skin of the thigh or upper arm is most effective because a larger portion can more easily be pinched.

* *Observation of Will Rice, summer 2010, age twelve, Lafayette, Louisiana.*

I2. Snake's Bite

Similar to Doctor's Shot, the Opies describe Snake's Bite as an "operation conducted merely by squeezing a piece of flesh between the thumbnails and rubbing in opposite directions."

* *See Opie and Opie (1959, 202).*

I3. Painful Pinkies

The actor inserts both pinkies into her mouth so that the pinky nails are directly beneath her bicuspids. She then bites down firmly with the points of the bicuspids in the center of the pinkie nail until just painful for a priming period of several seconds. The pinkies are then removed from her mouth and linked together in front of her sternum. The linked pinkies are then pulled against one another. The pain in the actor's pinkie nails "dramatically intensifies!"

* *Remembrance gathered from L. C. Baker, summer 2016, thirty-four, played at summer camp as a youth, Virginia.*

I4. George of the Jungle

The agent stands facing the patient. With his palms facing inward toward the patient's head, the agent begins to swing his arms rapidly past the patient's head, as if in a chopping motion. As the agent's arms fly, he sings the intro song from the 1960s animated television show *George of the Jungle*: "George, George, George of the Jungle / Strong as he could be / George, George, George of the Jungle / Watch out for that tree!" Upon uttering the final word, the agent smacks the patient on the forehead.

* *Remembrance gathered from John Anderson, 2010, thirty-eight, who played it as a teen growing up in Boise, Idaho.*

I4A. RUNNING THROUGH THE JUNGLE

Performance position and arm motions are the same as for I4. The agent states the patient is running through the jungle, dodging trees, turning left, then right, running past obstacles, then *stops!* The agent shouts the word *stop* and holds his palm up in front of the patient's face. The patient will suddenly stagger back.

* *Remembrance gathered from Jaron Nazaroff, summer 2016, twenty-four, who played it in his late teens in Santa Clara, California.*

J. Taste Illusions

J1. Matchstick Bacon

The actor lights and blows out a matchstick. The matchstick is then tasted. The taste is that of bacon.

* *Remembrance gathered from a folklore student, spring 2016.*

BIBLIOGRAPHY

adanatorememberful [pseud.]. 2012. "Vacuum—The Shitty School Parody [of] Fuck You." YouTube. October 25, 2012. Video, 3:14. https://www.youtube.com /watch?v=e8weiNbETB4.

American Folklore Society. n.d. "What is Folklore?" Accessed November 2, 2016. https://www.afsnet.org/?page=WhatIsFolklore.

Arbib, Michael A. 1987. "Schemas." In *The Oxford Companion to the Mind*, edited by Richard L. Gregory, 695–97. Oxford: Oxford University Press.

Armel, K. Carrie, and Vilayanur S. Ramachandran. 2003. "Projecting Sensations to External Objects: Evidence from Skin Conductance Response." *Proceedings of the Royal Society of London B: Biological Sciences* 270 (1523): 1499–1506.

Arnold, Harry L., Jr. 1952. "Japanese Illusion." *Science* 115 (2995): 577.

Artaud, Antonin. (1938) 1958. *The Theater and Its Double*. Translated by Mary Caroline Richards. New York: Grove.

Augustine. 2000. *Soliloquies: Augustine's Inner Dialogue*. Translated by Kim Paffenroth. Hyde Park, NY: New City.

Bakhtin, Mikhail. 1992. *The Dialogic Imagination: Four Essays*. Austin: University of Texas Press.

Barker, K. Brandon, and Claiborne Rice. 2012. "Folk Illusions: An Unrecognized Genre of Folklore." *Journal of American Folklore* 125 (498): 444–73.

———. 2016. "Folk Illusions and the Social Activation of Embodiment: Ping Pong, Olive Juice, and Elephant Shoe(s)." *Journal of Folklore Research* 53 (2): 63–85.

Barnhart, Anthony S. 2010. "The Exploitation by Gestalt Principles by Magicians." *Perception* 39 (9): 1286–89.

Bauman, Richard. (1977) 1984. *Verbal Art as Performance*. Prospect Heights, IL: Waveland.

———. 1982. "Ethnography of Children's Folklore." In *Children In and Out of School: Ethnography and Education*, edited by Perry Gilmore and Allan A. Glatthorn, 172–86. Language and Ethnography Series, vol. 2. Washington, DC: CAL.

Becker, Eni S., and Mike Rinck. 2004. "Sensitivity and Response Bias in Fear of Spiders." *Cognition and Emotion* 18 (7): 961–76.

Ben-Amos, Dan. 1969. "Analytical Categories and Ethnic Genres." *Genre* 2 (3): 275–301.

Benedetti, F. 1985. "Processing of Tactile Spatial Information with Crossed Fingers." *Journal of Experimental Psychology: Human Perception and Performance* 11 (4): 517–25.

———. 1986. "Tactile Diplopia (Diplesthesia) on the Human Fingers." *Perception* 15 (1): 83–91.

Bennett, Maxwell, Daniel Dennett, Peter Hacker, and John Searle. 2009. *Neuroscience and Philosophy: Brain, Mind, and Language*. New York: Columbia University Press.

Bergen, Benjamin K., and Nancy Chang. 2005. "Embodied Construction Grammar in Simulation-Based Language Understanding." In *Construction Grammars: Cognitive Grounding and Theoretical Extensions*, edited by Jan-Ola Östam and Mirjam Fried, 147–90. Philadelphia: John Benjamins.

Bertelson, Paul, and Beatrice De Gelder. 2004. "The Psychology of Multimodal Perception." In *Crossmodal Space and Crossmodal Attention*, edited by Charles Spence and Jon Driver, 141–77. Oxford: Oxford University Press.

Blakeslee, Sandra, and Matthew Blakeslee. 2007. *The Body Has a Mind of Its Own: How Body Maps in Your Brain Help You Do (Almost) Everything Better.* New York: Random House.

Boaz, Franz. (1928) 1986. *Anthropology and Modern Life.* New York: Dover.

Boulware, Jean T. 1951. "Numbness, Body-Image, and the Japanese Illusion." *Science* 114 (2970): 584–85.

Bower, T. G. R. 1966. "The Visual World of Infants." *Scientific American* 215 (6): 80–97.

———. 1971. "The Object in the World of the Infant." *Scientific American* 225 (4): 30–39.

Bronner, Simon J. 1988. *American Children's Folklore.* Little Rock, AR: August House.

Brugger, Peter, and Rebekka Meier. 2015. "A New Illusion at Your Elbow." *Perception* 44 (2): 219–21.

Bruner, Jerome, and V. Sherwood. 1976. "Peekaboo and the Learning of Rule Structure." In *Play: Its Role in Development and Evolution,* edited by Jerome Bruner, 277–85. New York: Basic Books.

Brunvand, Jan Harold. 1998. *The Study of American Folklore: An Introduction.* New York: Norton.

Buchanan, Daniel Crump. 1965. *Japanese Proverbs and Sayings.* Norman: University of Oklahoma Press.

Buckingham, Gavin. 2014. "Getting a Grip on Heaviness Perception: A Review of Weight Illusions and Their Probable Causes." *Experimental Brain Research* 232 (6): 1623–29.

Burnett, Charles Theodore. 1904. "Studies on the Influence of Abnormal Position upon the Motor Impulse." *Psychological Review* 11 (6): 371–94.

Caputo, Giovanni. 2010. "The Strange-Face-in-the-Mirror Illusion." *Perception* 39 (7): 1007–8.

Cavina-Pratesi, Cristiana, Gustav Kuhn, Magdalena Letswaart, and A. David Milner. 2011. "The Magic Grasp: Motor Expertise in Deception." *PLOS ONE* 6 (2): e16568. https://doi.org/10.1371/journal.pone.0016568.

Charlesworth, William R. 1966. "Persistence of Orienting and Attending Behavior in Infants as a Function of Stimulus-Locus Uncertainty." *Child Development* 37 (3): 473–91.

Charpentier, Augustin. 1891. "Analyse expérimentale de quelques éléments de la sensation de poids." *Archive de Physiologie Normale et Pathologique* 3:122–25.

Chica, Ana B., Elisa Martín-Arévalo, Fabiano Botta, and Juan Lupiáñez. 2014. "The Spatial Orienting Paradigm: How to Design and Interpret Spatial Attention Experiments." *Neuroscience and Biobehavioral Reviews* 40:35–51.

Chomsky, Noam. 2012. "I-Concepts, I-Beliefs, and I-Languages." In *The Science of Language: Interviews with James McGilvray,* 153–56. Cambridge: Cambridge University Press.

Cobb, Vicki. 1999. *How to Really Fool Yourself: Illusions for All Your Senses.* New York: Wiley.

Colman, Andrew M. 2015. *A Dictionary of Psychology.* Oxford: Oxford University Press.

Computer Science for Fun (CS4FN). n.d. "The Floating Arms Illusion." Computer Science for Fun. Accessed November 25, 2018. http://www.cs4fn.org/psychophysics/floatingarms.php.

Cook, Amy. 2006. "Staging Nothing: Hamlet and Cognitive Science." *SubStance* 35 (2): 83–99.

Cook, V. J., and Mark Newson. 2007. *Chomsky's Universal Grammar: An Introduction.* Malden, MA: Blackwell.

Cooper, John M. 1970. "Plato on Sense-Perception and Knowledge." *Phronesis* 15 (2): 123–46.

Coren, Stanley. 2013. "Sensation and Perception." In *Handbook of Psychology, V. 1 History of Psychology,* edited by Irving B. Wiener, 100–128. New York: John Wiley and Sons.

Craske, Brian, and J. D. Craske. 1986. "Oscillator Mechanisms in the Human Motor System: Investigating Their Properties Using the Aftercontraction Effect." *Journal of Motor Behavior* 18 (2): 117–45.

Csordas, Thomas. 1990. "Embodiment as Paradigm for Anthropology." *Ethos* 18 (1): 5–47.

———, ed. 1994a. *Embodiment and Experience: The Existential Ground of Culture and the Self*. Cambridge: Cambridge University Press.

———. 1994b. *The Sacred Self: A Cultural Phenomenology of Charismatic Healing*. Berkeley: University of California Press.

Czermak, Johann. 1855. "Physiologische Studien. III." Sitzungsberichte der Kaiserlichen Akademie der Wissenschaften (mathematisch-naturwissenschaftliche Classe) Vienna 17:563–600.

Damasio, Antonio. 1994. *Descartes' Error: Emotion, Reason, and the Human Brain*. New York: Putnam.

Darth Thanos [pseud.]. 2010. "Ping No Ball Pong." YouTube. July 12, 2010. Video, 0.50. http://www.youtube.com/watch?v=wg9Mo59H31I.

Davies, Owen. 2012. *Magic: A Very Short Introduction*. Oxford: Oxford University Press.

Daw, Nigel. 2005. *Visual Development*, 2nd edition. New York: Springer.

Dégh, Linda. 2001. *Legend and Belief*. Bloomington: Indiana University Press.

De Havas, Jack, Arko Ghosh, Hiroaki Gomi, and Patrick Haggard. 2015. "Sensorimotor Organization of a Sustained Involuntary Movement." *Frontiers in Behavioral Neuroscience* 9 (article 185): 1–15.

De Havas, Jack, Hiroaki Gomi, and Patrick Haggard. 2017. "Experimental Investigations of Control Principles of Involuntary Movement: A Comprehensive Review of the Kohnstamm Phenomenon." *Experimental Brain Research* 235:1953–97.

———. 2016. "Voluntary Motor Commands Reveal Awareness and Control of Involuntary Movement." *Cognition* 155:155–67.

Derbyshire, Stuart W. G., Matthew G. Whalley, V. Andrew Stenger, and David A. Oakley. 2004. "Cerebral Activation during Hypnotically Induced and Imagined Pain." *Neuroimage* 23 (1): 392–401.

Deręgowski, Jan B. 2013. "On the Muller-Lyer Illusion in the Carpentered World." *Perception* 42:790–92.

Dieguez, Sebastian, Manuel R. Mercier, Nate Newby, and Olaf Blanke. 2009. "Feeling Numbness for Someone Else's Finger." *Current Biology* 19 (24): R1108–R1109.

Doniger, Wendy. 1998. *The Implied Spider: Politics and Theology in Myth*. New York: Columbia University Press.

———. 2009. *The Hindus: An Alternative History*. New York: Penguin.

Dormtainment [pseud.]. 2011. "Penny Pong - @Dormtainment." YouTube. September 24, 2011. Video, 4:42. https://www.youtube.com/watch?v=7ykXK2FC2fM.

Doyle, Arthur Conan. 1976. *The Illustrated Sherlock Holmes Treasury*. New York: Avenel/Crown.

Duclos, Cyril, Régine Roll, Anne Kavounoudias, and Jean-Pierre Roll. 2004. "Long-Lasting Body Leanings Following Neck Muscle Isometric Contractions." *Experimental Brain Research* 158 (1): 58–66.

———. 2007. "Cerebral Correlates of the 'Kohnstamm Phenomenon': An fMRI Study." *NeuroImage* 34:774–83.

Dolezal, Luna. 2015. *The Body and Shame: Phenomenology, Feminism, and the Socially Shaped Body*. Lanham, MD: Lexington.

Dreyfus, Hubert L. 1996. "The Current Relevance of Merleau-Ponty's Phenomenology of Embodiment." *Electronic Journal of Analytic Philosophy* 4 (Spring). http://ejap.louisiana.edu/EJAP/1996.spring/dreyfus.1996.spring.html.

Dundes, Alan. 1971. "Folk Ideas as Units of Worldview." *Journal of American Folklore* 84 (331): 93–103.

———. 1972. "Seeing Is Believing." *Natural History Magazine* 81 (5): 8–12, 86–87.

———, ed. 1984. *Sacred Narrative: Readings in the Theory of Myth*. Berkeley: University of California Press.

———. 1997. "The Motif-Index and the Tale-Type Index: A Critique." *Journal of Folklore Research* 34 (3): 195–202.

———. 1998. "Bloody Mary in the Mirror: A Ritual Reflection of Pre-pubescent Anxiety." *Western Folklore* 57 (2/3): 119–35.

During, Simon. 2002. *Modern Enchantments: The Cultural Power of Secular Magic*. Cambridge, MA: Harvard University Press.

Eimas, Peter D., Einar R. Siqueland, Peter Juscyzk, and James Vigorito. 1971. "Speech Perception in Infants." *Science* 171 (3968): 303–6.

Ellis, Robert R., and Susan J. Lederman. 1998. "The Golf-Ball Illusion: Evidence for Top-Down Processing in Weight Perception." *Perception* 27 (2): 193–201.

Enfield, N. J., Asifa Majid, and Miriam van Staden. 2006. "Cross-Linguistic Categorisation of the Body: Introduction." *Language Sciences* 28 (2–3): 137–47.

EvanCG Productions [pseud.]. 2014. "Sousaphone Lightsaber Battle." YouTube. September 14, 2014. Video, 0:13. https://www.youtube.com/watch?v=4tyJSxB1Rdc.

Fales, Cornelia. 2002. "The Paradox of Timbre." *Ethnomusicology* 46 (1): 56–95.

Fauconnier, Gilles. 2001. "Conceptual Blending and Analogy." In *The Analogical Mind: Perspectives from Cognitive Science*, edited by Dedre Gentner, Keith J. Holyoak, and Boicho N. Kokinov, 255–85. Cambridge, MA: MIT Press.

Fauconnier, Gilles, and Mark Turner. (2002) 2008. *The Way We Think: Conceptual Blending and the Mind's Hidden Complexities*. New York: Basic Books.

Fernald, Anne, and Daniela K. O'Neill. 1993. "Peekaboo across Cultures: How Mothers and Infants Play with Voices, Faces, and Expectations." In *Parent-Child Play: Descriptions and Implications*, edited by Kevin MacDonald, 259–85. Albany: State University of New York Press.

Finnegan, Ruth. (1970) 2012. *Oral Literature in Africa*. Cambridge: Open Book. http://dx.doi.org/10.11647/OBP.0025.

Flanagan, J. Randall, and Michael A. Beltzner. 2000. "Independence of Perceptual and Sensorimotor Predictions in the Size-Weight Illusion." *Nature Neuroscience* 3 (7): 737–41.

Flanagan, J. Randall, Sara King, Daniel M. Wolpert, and Roland S. Johansson. 2001. "Sensorimotor Prediction and Memory in Object Manipulation." *Canadian Journal of Experimental Psychology* 55 (2): 89–97.

Frazer, James G. (1890) 1951. *The Golden Bough: A Study in Magic and Religion*. New York: Macmillan.

Furman, Wyndol, B. Bradford Brown, and Candice Feiring, eds. 1999. *The Development of Romantic Relationships in Adolescence*. Cambridge: Cambridge University Press.

Gallop, David. 1990. *Aristotle on Sleep and Dreams: A Text and Translation with Introduction, Notes and Glossary*. Ontario: Broadview.

Geertz, Clifford. 1980. "Blurred Genres: The Reconfiguration of Social Thought." *American Scholar* 49 (2): 165–79.

Geldard, Frank A., and Carl E. Sherrick. "The Cutaneous 'Rabbit': A Perceptual Illusion." *Science* 178 (4057): 178–79.

Ghosh, Arko, John Rothwell, and Patrick Haggard. 2014. "Using Voluntary Motor Commands to Inhibit Involuntary Arm Movements." *Proceedings of the Royal Society B* 281 (1794): 20141139.

Gibbs, Raymond, Jr. 2005. *Embodiment and Cognitive Science*. Cambridge: Cambridge University Press.

———. 2017. *Metaphor Wars: Conceptual Metaphors in Human Life*. Cambridge: Cambridge University Press.

Gibson, Eleanor Jack, and Anne D. Pick. 2000. *An Ecological Approach to Perceptual Learning and Development*. Oxford: Oxford University Press.

Gibson, James J. (1977) 2015. *The Ecological Approach to Visual Perception*. New York: Psychology Press.

Gilbert, George, and Wendy Rydell. 1976. *Great Tricks of the Master Magicians, with Illustrated Instructions for Performing Them and Many Others*. New York: Golden Press.

Glassie, Henry. (1975) 1983. *All Silver and No Brass: An Irish Christmas Mumming*. Philadelphia: University of Pennsylvania Press.

———. 1989. *The Spirit of Folk Art*. New York: Henry N. Abrams and Museum of International Folk Art.

Glassie, Henry, and Pravina Shukla. 2018. *Sacred Art: Catholic Saints and Candomblé Gods in Modern Brazil*. Bloomington: Indiana University Press.

Goldin-Meadow, Susan. 2003. *The Resilience of Language: What Gesture Creation in Deaf Children Can Tell Us about How All Children Learn Language*. New York: Psychology Press.

Goldsmith, H. H., and M. K. Rothbart. 1999. "The Laboratory Temperament Assessment Battery (Lab-TAB): Pre-Locomotor Version 3.1." University of Oregon, Department of Psychology.

Goldstein, Diane. 2007. "Scientific Rationalism and Supernatural Experience Narratives." In *Haunting Experiences: Ghosts in Contemporary Folklore*, edited by Diane Goldstein, Sylvia Ann Grider, and Jeannie Banks Thomas, 60–78. Logan: Utah State University Press.

Goldstein, Kenneth. 1967. "The Induced Natural Context: An Ethnographic Field Technique." In *Essays in the Verbal and Visual Arts*, edited by June Helm, 1–6. Seattle: University of Washington Press.

Goodwin, Charles. 2000. "Action and Embodiment within Situated Human Interaction." *Journal of Pragmatics* 32:1489–1522.

Gould, Stephen Jay. 1997. "The Exaptive Excellence of Spandrels as a Term and Prototype." *Proceedings of the National Academy of Sciences* 94 (20): 10750–55.

Gould, Stephen Jay, and R. C. Lewontin. 1979. "The Spandrels of San Marco and the Panglossian Paradigm: A Critique of the Adaptationist Programme." *Proceedings of the Royal Society of London B: Biological Sciences* 205 (1161): 581–98.

Greenfield, P. M. 1972. "Playing Peekaboo with a Four-Month-Old: A Study of the Role of Speech and Nonspeech Sounds in the Formation of a Visual Schema." *Journal of Psychology* 82 (2): 287–98.

Gregory, Richard L. 1973. "The Confounded Eye." In *Illusion in Nature and Art*, edited by E. H. J. Gombrich and Richard L. Gregory, 49–95. New York: Scribner.

———. 1998. *Mirrors in Mind*. London: Penguin.

———. 2009. *Seeing through Illusions*. Oxford: Oxford University Press.

———. 2012. "Pictures as Strange Objects of Perception." In *Sensory Perception: Mind and Matter*, edited by Friedrich G. Barth, Patrizia Giampieri-Deutsch, and Hans-Dieter Klein, 175–81. New York: Springer.

Grider, Sylvia Ann. 2007. "Children's Ghost Stories." In *Haunting Experiences: Ghosts in Contemporary Folklore*, edited by Diane Goldstein, Sylvia Ann Grider, and Jeannie Banks Thomas, 111–40. Logan: Utah State University Press.

Hall, G. Stanley, and Arthur Allin. 1897. "The Psychology of Tickling, Laughing, and the Comic." *American Journal of Psychology* 9 (1): 1–41.

Hammond, Joyce D. 1995. "The Tourist Folklore of Pele: Encounters with the Other." In *Out of the Ordinary: Folklore and the Supernatural*, edited by Barbra Walker, 159–79. Logan: Utah State University Press.

Hanna-Attisha, Mona, Jenny LaChance, Richard Casey Sadler, and Allison Champney Schnepp. 2016. "Elevated Blood Lead Levels in Children Associated with the Flint Drinking Water Crisis: A Spatial Analysis of Risk and Public Health Response." *American Journal of Public Health* 106 (2): 283–290.

Henri, Victor. 1898. *Über die Raumwahrnehmungen des Tastsinnes: Ein Beitrag zur experimentellen Psychologie*. Berlin: Verlag von Reuther and Reichard.

Hogan, Patrick Colm. 2003. *The Mind and Its Stories: Narrative Universals and Human Emotion*. Cambridge: Cambridge University Press.

Hudson, Arthur Palmer, and Pete Kyle McCarter. 1934. "The Bell Witch of Tennessee and Mississippi: A Folk Legend." *Journal of American Folklore* 47 (183): 45–63.

Huizinga, Johan. (1944) 1949. *Homo Ludens: A Study of the Play-Element in Culture*. London: Routledge and Kegan Paul.

Hymes, Dell. (1974) 1980. *Foundations in Sociolinguistics: An Ethnographic Approach*. Philadelphia: University of Pennsylvania Press.

———. 1975a. "Breakthrough into Performance." In *Folklore: Performance and Communication*, edited by Dan Ben-Amos and Kenneth S. Goldstein, 11–74. The Hague: Mouton.

———. 1975b. "Folklore's Nature and the Sun Myth." *Journal of American Folklore* 88 (350): 345–69.

IMDB (The Internet Movie Database). n.d. "Results for 'Bloody Mary.'" Accessed November 25, 2018. https://www.imdb.com/find?ref_=nv_sr_fn&q=bloody+mary&s=all.

Ivanenko, Yuri P., W. Geoffrey Wright, Victor S. Gurfinkel, Fay Horak, and Paul Cordo. 2006. "Interaction of Involuntary Post-Contraction Activity with Locomotor Movements." *Experimental Brain Research* 169 (2): 255–60.

Jackendoff, Ray. 1987. "The Status of Thematic Relations in Linguistic Theory." *Linguistic Inquiry* 18 (3): 369–411.

Janeček, Petr. 2014. "Bloody Mary or Krvavá Mary? Globalization and Czech Children's Folklore." *Slovak Ethnology/Slovensky Narodopis* 62 (2): 221–43.

Johannsen, Dorothea E. 1971. "Early History of Perceptual Illusions." *Journal of the History of the Behavioral Sciences* 7 (2): 127–40.

Johnson, Mark. (1987) 1990. *The Body in the Mind: The Bodily Basis of Meaning, Imagination, and Reason*. Chicago: University of Chicago Press.

Jones, Michael Owen. 1989. *Craftsman of the Cumberlands: Tradition and Creativity*. Lexington: University of Kentucky Press.

Kidzworld. n.d. "Body and Mind Experiments." Accessed August 15, 2011. http://www
.kidzworld.com/article/1917-body-and-mind-experiments.

Klein, Elmer, and Paul Schilder. 1929. "The Japanese Illusion and the Postural Model of the
Body." *Journal of Nervous and Mental Disease* 70 (3): 241–63.

Knapp, Mary, and Herbert Knapp. 1976. *One Potato, Two Potato . . . : The Folklore of
American Children.* New York: Norton.

Koch, Christof, and Tomaso Poggio. 1999. "Predicting the Visual World: Silence Is Golden."
Nature Neuroscience 2 (1): 9–10.

Kohnstamm, Oskar. 1915. "Demonstration einer Katatonieartigen Erscheinung beim
Gesunden (Katatonusversuch)." *Neurologisches Centralblatt* 34 (9): 290–91.

Kuhn, Gustav, Alym A. Amlani, and Ronald A. Rensink. 2008. "Towards a Science of Magic."
Trends in Cognitive Science 12 (9): 349–54.

Kuhn, Gustav, and Michael F. Land. 2006. "There's More to Magic Than Meets the Eye."
Current Biology 16 (22): 950–51.

Lakoff, George, and Mark Johnson. 1999. *Philosophy in the Flesh: The Embodied Mind and Its
Challenge to Western Thought.* New York: Basic Books.

Langacker, Ronald W. 1991. *Foundations of Cognitive Grammar. Vol. 2: Descriptive
Application.* Stanford, CA: Stanford University Press.

Langlois, Janet. (1978) 1980. "'Mary Whales I Believe in You': Myth and Ritual Subdued." In
Indiana Folklore: A Reader, edited by Linda Dégh, 196–224. Bloomington: Indiana
University Press.

Leder, Drew. 1990. *The Absent Body.* Chicago: University of Chicago Press.

Lloyd, G. E. R. 1991. "Observational Error in Later Greek Science." In *Methods and
Problems in Greek Science: Selected Papers,* 299–332. Cambridge: Cambridge University
Press.

Lockhart, Teresa Ann Bell. (1984) 2009. "Twentieth-Century Aspects of the Bell Witch."
In *A Tennessee Folklore Sampler: Selections from the Tennessee Folklore Society
Bulletin, 1935–2009,* edited by Ted Olson and Anthony P. Cavender, 234–42. Knoxville:
University of Tennessee Press.

Logothetis, Nikos K. 2008. "What We Can Do and What We Cannot Do with FMRI." *Nature*
453 (7197): 869–78.

LoloYodel [pseud.]. 2011. "Garcia & Weir on Letterman 9-17-1987, New York, NY (LoloYodel)."
YouTube. August 14, 2011. Video, 14:51. https://youtu.be/LH6nKv5SVFQ.

Macknik, Stephen L., Mac King, James Randi, Apollo Robbins, John Thompson, and Susana
Martinez-Conde. 2008. "Attention and Awareness in Stage Magic: Turning Tricks into
Research." *Nature Reviews. Neuroscience* 9 (11): 871–79.

Macknik, Stephen L., and Susana Martinez-Conde. 2010. *Sleights of Mind: What the
Neuroscience of Magic Reveals about Our Everyday Deceptions.* New York: Picador.

Maclachlan, Malcolm. 2004. *Embodiment: Clinical, Critical, and Cultural Perspectives on
Health and Illness.* New York: Open University Press.

Mauss, Marcel. (1950) 1972. *A General Theory of Magic.* Translated by Robert Brain. New
York: Norton.

McDowell, John. 1999. "The Transmission of Children's Folklore." In *Children's Folklore: A
Sourcebook,* edited by Brian Sutton-Smith, Jay Mechling, Thomas W. Johnson, and
Felicia R. McMahon, 49–62. Logan: Utah State University Press.

McGurk, Harry, and John MacDonald. 1976. "Hearing Lips and Seeing Voices." *Nature* 264
(5588): 746–48.

Melville, Herman. (1851) 1992. *Moby Dick*. Hertfordshire: Wordsworth Classics.

Merleau-Ponty, Maurice. 1962. *Phenomenology of Perception*. Translated by Colin Smith. London: Routledge and Kegan Paul.

Mol, Annemarie. 2008. *The Logic of Care: Health and the Problem of Patient Choice*. New York: Routledge.

Montague, Diane P. F., and Arlene S. Walker-Andrews. 2001. "Peekaboo: A New Look at Infants' Perception of Emotion Expressions." *Developmental Psychology* 37 (6): 826–38.

Nakayama, Ken. 2005. "Modularity in Perception, Its Relation to Cognition and Knowledge." In *Blackwell Handbook of Sensation and Perception*, edited by E Bruce Goldstein, 737–59. New York: Blackwell.

Nash, Michael R., and Amanda J. Barnier, eds. 2012. *The Oxford Handbook of Hypnosis: Theory, Research, and Practice*. Oxford: Oxford University Press.

Ninio, Jacques. (1998) 2001. *The Science of Illusions*. Ithaca, NY: Cornell University Press.

———. 2014. "Geometric Illusions Are Not Always Where You Think They Are: A Review of Some Classical and Less Classical Illusions and Ways to Describe Them." *Frontiers in Human Neuroscience* 8 (article 856). https://www.frontiersin.org/articles/10.3389/fnhum.2014.00856.

Nisbett, Richard E., and Timothy D. Wilson. 1977. "Telling More Than We Can Know: Verbal Reports on Mental Processes." *Psychological Review* 84 (3): 231–59.

Noyes, Dorothy. 2012. "The Social Base of Folklore." In *A Companion to Folklore*, edited by Regina F. Bendix and Galit Hasan-Rokem, 13–39. Malden, MA: Wiley-Blackwell.

Oakley, David A., and Peter W. Halligan. 2013. "Hypnotic Suggestion: Opportunities for Cognitive Neuroscience." *Nature Reviews Neuroscience* 14 (8): 565–76.

O'Connor, Kathleen Malone. 2010. "Magic." In *Folklore: An Encyclopedia of Beliefs, Customs, Tales, Music, and Art*, 3 vols., edited by Charlie T. McCormick and Kim Kennedy White, 811–21. Santa-Barbara, CA: ABC-CLIO.

Okada, Kayoko, and Gregory Hickok. 2009. "Two Cortical Mechanisms Support the Integration of Visual and Auditory Speech: A Hypothesis and Preliminary Data." *Neuroscience Letters* 452 (3): 219–23.

Opie, Iona, and Peter Opie. 1959. *The Lore and Language of Schoolchildren*. Oxford: Clarendon.

Oring, Elliott. 2012. *Just Folklore: Analysis, Interpretation, Critique*. Los Angeles: Cantilever.

Otte, Ellen, and Hanneke I. van Mier. 2006. "Bimanual Interference in Children Performing a Dual Motor Task." *Human Movement Science* 25 (4): 678–93.

Parrott, W. Gerrod, and Henry Gleitman. 1989. "Infant's Expectations in Play: The Joy of Peekaboo." *Cognition and Emotion* 3 (4): 291–311.

Pasquinelli, Elena. 2012. "The Awareness of Illusions." In *Perceptual Illusion: Philosophical and Psychological Essays*, edited by Clotilde Calabi, 59–74. New York: Palgrave Macmillan.

Pereira, Jayme R. 1925. "Contraction automatique des muscles stries chez l'homme." *Journal de Physiologie et de Pathologie Générale* 23 (1): 30–38.

Petitto, Laura-Ann, and Paula F Marentette. 1991. "Babbling in the Manual Mode: Evidence for the Ontogeny of Language." *Science* 251 (5000): 1483–96.

Petroski, Henry. 1992. *The Pencil: A History of Design and Circumstance*. New York: Knopf.

Piaget, Jean. (1969) 2013. *The Mechanisms of Perception*. New York: Routledge.

Pomerantz, James. 1983. "The Rubber Pencil Illusion." *Perception and Psychophysics* 33 (4): 365–68.

Povinelli, Daniel J. 2012. *World without Weight: Perspectives on an Alien Mind*. Oxford: Oxford University Press.

Povinelli, Daniel J., Keli R. Landau, and Helen K. Perilloux. 1996. "Self-Reflection in Young Children Using Delayed Versus Live Feedback: Evidence of a Developmental Asynchrony." *Child Development* 67 (4): 1540–54.

Ramachandran, Vilayanur S., and Sandra Blakeslee. 1999. *Phantoms in the Brain: Probing the Mysteries of the Human Mind*. New York: William Morrow.

Ramachandran, Vilayanur S., Beatrix Krause, and Laura K. Case. 2011. "The Phantom Head." *Perception* 40 (3): 367–70.

Ramachandran, Vilayanur S., and Diane Rogers-Ramachandran. 2008. "Touching Illusions." *Scientific American* 18 (2): 60–63.

Ramachandran, Vilayanur S., Diane Rogers-Ramachandran, and Steve Cobb. 1995. "Touching the Phantom Limb." *Nature* 377 (6549): 489.

Randi, James. 1992. *Conjuring*. New York: St. Martin's Press.

Ratner, Nancy, and Jerome Bruner. 1978. "Games, Social Exchange, and the Acquisition of Language." *Journal of Child Language* 5 (3): 391–401.

Richmond, W. Edson. 1986. Introduction to *Handbook of American Folklore* by Richard Dorson. Bloomington: Indiana University Press.

Rickels, Patricia. 1961. "Some Accounts of Witch Riding." *Louisiana Folklore Miscellany* 2 (1): 53–63.

Rivers, W. H. R. 1905. "Observation on the Senses of the Todas." *British Journal of Psychology* 1 (4): 321–96.

Roberts, Katharine. 1998. "Contemporary Cauchemar: Experience, Belief, and Prevention." *Louisiana Folklore Miscellany* 8:15–26.

Robinson, James O. (1972) 1998. *The Psychology of Visual Illusion*. New York: Dover.

Rohault, Jacques. (1671) 1723. *Rohault's System of Natural Philosophy, Illustrated with Dr. Samuel Clarke's Notes: Taken Mostly out of Sir Isaac Newton's Philosophy with Additions*. Translated by John Clarke. London: J. Knapton.

Rosenblum, Lawrence D. 2010. *See What I'm Saying: The Extraordinary Powers of Our Five Senses*. New York: Norton.

Ross, W. D. 1924. *Aristotle's Metaphysics: A Revised Text with Introduction and Commentary*. The Internet Classics Archive. Accessed November 2, 2016. http://classics.mit.edu /Aristotle/metaphysics.4.iv.html.

Rydell, Wendy, and George Gilbert. 1977. *The Great Book of Magic: Including 150 Mystifying Tricks You Can Perform*. New York: Abrams.

Schaub, Simone, Evelyn Bertin, and Trix Cacchione. 2013. "Infants' Individuation of Rigid and Plastic Objects Based on Shape." *Infancy* 18 (4): 628–38.

Schilder, Paul. 1935. *The Image and Appearance of the Human Body*. London: Kegan.

Schoneberger, Ted. 2010. "Three Myths from the Language Acquisition Literature." *The Analysis of Verbal Behavior* 26 (1): 107–31.

Schwartz, A., and P. Meyer. 1921. "Un curieux phénomène d'automatisme chez l'homme." *Comptes Rendus de la Société de Biologie* 85 (27): 490–92.

Schwarz, Luis, and Stanton Fjeld. 1968. "Illusions Induced by the Self-Reflected Image." *Journal of Nervous and Mental Disease* 146 (4): 277–84.

Schrempp, Gregory. 1996. "Folklore and Science: Inflections of 'Folk' in Cognitive Research." *Journal of Folklore Research* 33 (3): 191–206.

———. 2012. *Ancient Mythology of Modern Science: A Mythologist Looks (Seriously) at Popular Science Writing*. Montreal: McGill-Queen's University Press.

Segall, Marshall H., Donald T. Campbell, and Melville J. Herskovits. 1963. "Cultural Differences in the Perception of Geometric Illusions." *Science* 139 (3556): 769–71.

Selden, Samuel T. 2004. "Tickle." *Journal of the American Academy of Dermatology* 50 (1): 93–97.

Shakespeare, William. 2000. *Merchant of Venice*. In *The Illustrated Stratford Shakespeare*, 190–213. London: Chancellor Press.

Shams, Ladan, Yukiyasu Kamitani, and Shinsuke Shimojo. 2000. "Illusions: What You See Is What You Hear." *Nature* 408 (6814): 788.

Sklar, Deidre. 1994. "Can Bodylore Be Brought to Its Senses?" *Journal of American Folklore* 107 (423): 9–22.

———. 2005. "The Footfall of Words: A Reverie on Walking with the Nuestra Senora de Guadalupe." *Journal of American Folklore* 118 (467): 9–20.

Spence, Charles, Francesco Pavani, Angelo Maravita, and Nicholas Holmes. 2004. "Multisensory Contributions to the 3-D Representation of Visuotactile Peripersonal Space in Humans: Evidence from the Crossmodal Congruency Task." *Journal of Physiology* 98 (1): 171–89.

Steve Lamar [pseud.]. 2011. "No Ball Plate Ping Pong." YouTube. September 17, 2011. Video, 0.23. http://www.youtube.com/watch?v=9Cv3l9XSGi4.

Stewart, Susan. 1991. "Notes on Distressed Genres." *Journal of American Folklore* 104 (411): 5–31.

Sutton, David E. 2001. *Remembrance of Repasts: An Anthropology of Food and Memory*. New York: Bloomsbury.

Sutton-Smith, Brian. 1970. "Psychology of Childlore: The Triviality Barrier." *Western Folklore* 29 (1): 1–8.

———. 1973. "Games, the Socialization of Conflict." *Canadian Journal of History of Sport and Physical Education* 4 (1): 1–7.

———. 1981. *The Folkstories of Children*. Philadelphia: University of Pennsylvania Press.

———. (1997) 2001. *The Ambiguity of Play*. Cambridge, MA: Harvard University Press.

———. 1999. "What Is Children's Folklore?" In *Children's Folklore: A Sourcebook*, edited by Brian Sutton-Smith, Jay Mechling, Thomas W. Johnson, and Felicia R. MacMahon, 3–9. Logan: Utah State University Press.

Thaler, Lore, James T. Todd, Miriam Spering, and Karl R. Gegenfurtner. 2007. "Illusory Bending of a Rigidly Moving Line Segment: Effects of Image Motion and Smooth Pursuit Eye Movements." *Journal of Vision* 7 (6): 1–13.

Thompson, Stith. (1955–58) 2000. *Motif-Index of Folk Literature*. Electronic edition. Charlottesville, VA: Intelex Corporation.

Tomasello, Michael. 1999. *The Cultural Origins of Human Cognition*. Cambridge, MA: Harvard University Press.

Tremaine, Jon. 2002. *Magic and Card Tricks*. Bath, UK: Parragon.

Tucker, Elizabeth. 1984. "Levitation and Trance Sessions at Preadolescent Girls' Slumber Parties." In *The Masks of Play*, edited by Brian Sutton-Smith and Diana Kelly-Byrne, 125–33. New York: Leisure.

———. 2005. "Ghosts in Mirrors: Reflections of Self." *Journal of American Folklore* 118 (468): 186–203.

———. 2007. "Levitation Revisited." *Children's Folklore Review* 30:47–60.

———. 2008. *Children's Folklore: A Handbook.* Westport, CT: Greenwood.

———. 2012. "Changing Concepts of Childhood: Children's Folklore Scholarship since the Late Nineteenth Century." *Journal of American Folklore* 125 (3): 389–410.

Tylor, Edward Burnett. 1871. *Primitive Culture: Researches into the Development of Mythology, Philosophy, Religion, Art, and Custom.* Vol. 2. London: J. Murray.

Van Le, Quan, Lynne A. Isbell, Jumpei Matsumoto, Minh Nguyen, Etsuro Hori, Rafael S. Maior, Carlos Tomaz, Anh Hai Tran, Taketoshi Ono, and Hisao Nishijo. 2013. "Pulvinar Neurons Reveal Neurobiological Evidence of Past Selection for Rapid Detection of Snakes." *Proceedings of the National Academy of Sciences of the United States of America* 110 (47): 19000–19005. https://doi.org/10.1073/pnas.1312648110.

Van Riper, Charles. 1935. "An Experimental Study of the Japanese Illusion." *American Journal of Psychology* 47 (2): 252–63.

Vida, Mark D., and Daphne Maurer. 2012. "The Development of Fine-Grained Sensitivity to Eye Contact after 6 Years of Age." *Journal of Experimental Child Psychology* 112 (2): 243–56.

von Helmholtz, Hermann. (1910) 2005. *Treatise on Physiological Optics, Volume III.* Edited by James P. C. Southall. New York: Dover.

Washburn, R. W. 1929. "A Study of the Smiling and Laughing of Infants in the First Year of Life." *Genetic Psychology Monographs* 6 (5): 403–537.

Welsch, Roger. 1966. "Nebraska Finger Games." *Western Folklore* 25 (3): 173–94.

Winkelman, Michael, et al. 1982. "Magic: A Theoretical Reassessment [and Comments and Replies]." *Current Anthropology* 23 (1): 37–66.

Wilson, Frank R. 1998. *The Hand: How Its Use Shapes the Brain, Language, and Culture.* New York: Pantheon.

Witkowski, Stanley R., and Cecil H. Brown. 1985. "Climate, Clothing, and Body-Part Nomenclature." *Ethnology* 24 (3): 197–214.

Yang, Quing, and Zoi Kapoula. 2004. "Saccade-Vergence Dynamics and Interaction in Children and in Adults." *Experimental Brain Research* 156 (2): 212–23.

Young, J. Jay, Hong Z. Tan, and Rob Gray. 2003. "Validity of Haptic Cues and Its Effect on Priming Visual Spatial Attention." In *Haptic Interfaces for Virtual Environment and Teleoperator Systems, 2003. HAPTICS 2003. Proceedings. 11th Symposium,* 166–70.

Young, Katherine, ed. 1993. *Bodylore.* Knoxville: University of Tennessee Press.

———. 2011. "Gestures, Intercorporeity, and the Fate of Phenomenology in Folklore." *Journal of American Folklore* 124 (492): 55–87.

Zeiler, Kristin. 2010. "A Phenomenological Analysis of Bodily Self-Awareness in the Experience of Pain and Pleasure: On *Dys*-Appearance and *Eu*-Appearance." *Medicine, Health Care, and Philosophy* 13 (4): 333–42.

Zeman, Astrid, Oliver Obst, and Kevin R. Brooks. 2014. "Complex Cells Decrease Errors for the Müller-Lyer Illusion in a Model of the Visual Ventral Stream." *Frontiers in Computational Neuroscience* 8 (article 112): 1–9.

INDEX OF SUBJECTS

239

INDEX OF NAMES

INDEX OF FOLK ILLUSIONS

K. BRANDON BARKER is Lecturer in Folklore at Indiana University Bloomington. This is his first book.

CLAIBORNE RICE is Associate Professor of English at the University of Louisiana at Lafayette. This is his first book.

CPSIA information can be obtained
at www.ICGtesting.com
Printed in the USA
LVHW080150220421
685119LV00033BA/609